Related volumes
Brickwork 1 and Associated Studies } Harold Bailey and David Hancock
Brickwork 3 and Associated Studies

Other title of interest
The Skills of Plastering, Mel Baker

BRICKWORK 2
AND ASSOCIATED STUDIES

Harold Bailey

Sometime Senior Lecturer
Stockport College of Technology

David Hancock

Senior Lecturer
Stockport College of Technology

Second Edition

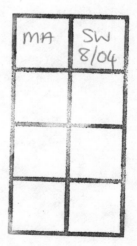

MACMILLAN

First edition 1979
Reprinted 1986, 1987, 1988
Second edition 1990

Published by
MACMILLAN EDUCATION LTD
Houndmills, Basingstoke, Hampshire RG21 2XS
and London
Companies and representatives
throughout the world

Printed in Hong Kong

British Library Cataloguing in Publication Data
Bailey, Harold
    Brickwork.——2nd ed
    2
    1. Brickwork——Manuals
    I. Title   II. Hancock, David W.
    693'.21

ISBN 0-333-51956-6

CONTENTS

PREFACE

This series of three volumes is designed to provide an introduction to the brickwork craft and the construction industry for craft apprentices and all students involved in building. All too often, new entrants to the construction industry are expected to have a knowledge of calculations, geometry, science and technology irrespective of their previous education. It is the authors' aim to provide a course of study which is not only easily understood but is also able to show the relationship that exists between technology and associated studies.

The construction industry recognises that the modern craftsman, while maintaining a very high standard of skills, must be capable of accepting change — in methods, techniques and materials. Therefore it will be necessary for apprentices to develop new skills related to the constant advancements in technology.

This second volume contains the basic knowledge necessary for the creative skills to be mastered by the apprentice. All elements contained within the superstructure are fully described and illustrated to the standard now required in modern building practice. Each chapter covers all the necessary information for the apprentice studying for examinations, and also for the improvement of his knowledge and understanding.

The authors, recognising the necessity for an improvement in efficiency and productivity, suggest that the apprentice should always have these factors as his objectives throughout his career in the construction industry.

HAROLD BAILEY
DAVID HANCOCK

ACKNOWLEDGEMENTS

The authors wish to acknowledge the assistance and cooperation of: Allmat Ltd, for figures 4.10 and 4.11 from promotional material for Furfix, Wall Extension Profiles; Forticrete Ltd, for figures 7.24 and 7.25; The Solid Fuel Advisory Service, for figures 11.4, 11.13, 11.15, 11.16, 11.20, 11.21, 11.22 and 11.23; Kopex Ltd, for figure 11.27; The Cement & Concrete Association, for figure 12.1.

Every effort has been made to trace all the copyright holders, but if any have been inadvertently overlooked the publisher will be pleased to make the necessary arrangement at the first opportunity.

FOREWORD: SAFETY IN BUILDING

There were almost 20,000 reportable injuries (those injuries involving more than three days' absence from work) in the Building Industry between April 1987 and April 1988, including over 150 deaths.

The authors appreciate this opportunity to bring these appalling figures to the immediate attention of apprentice bricklayers, and at the same time to remind them of their responsibilities, as well as those of their employers.

The *Health and Safety at Work Act 1974* became effective in 1982. This Act made further provision to the existing *Construction Regulations* for ensuring the health, safety and welfare of persons at work, and may be briefly summarised as follows:

An employer must ensure as far as practicable
1. The health, safety and welfare of his employees while at work.
2. The provision, and maintenance of safe plant and systems of work.
3. Information, instruction, training and supervision as necessary.
4. A safe place of work.

While at work an employee must
1. Take reasonable care of the health and safety of himself and all other persons who may be affected by his acts or omissions.
2. Co-operate fully with management in all health and safety matters.
3. Not interfere with, or misuse anything provided in the interests of health and safety.

1
SOLID GROUND FLOORS

Solid floors have a long history, dating back as far as the Roman Empire. They were used by mediaeval builders, the usual finish being of natural stone flags; in the nineteenth century many small cottages had ground floors consisting of a lime—ash mixture, trowelled to a smooth finish. Modern ground floors are generally considered to result from the shortage of timber after the Second World War.

Basically, a solid floor is a concrete slab laid on a base of suitable hardcore, which has been placed and consolidated after a layer of vegetable soil has been removed. At some level a damp-proof membrane (d.p.m.) must be incorporated and the slab is normally covered with a perfectly level, smooth sand-and-cement screed to receive a tiled finish. The usual build-up on site for the construction of a solid floor is shown in figure 1.1; the damp-proof membrane and the screed are normally completed at a later date.

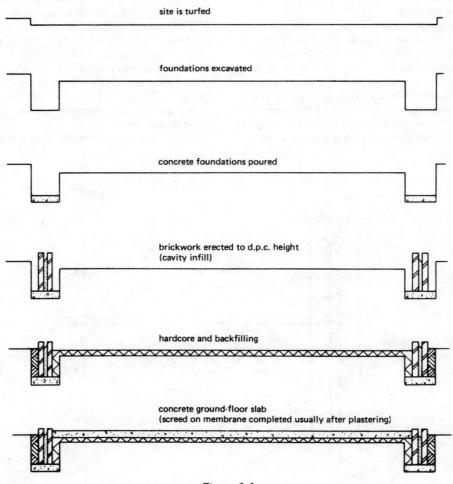

site is turfed

foundations excavated

concrete foundations poured

brickwork erected to d.p.c. height
(cavity infill)

hardcore and backfilling

concrete ground-floor slab
(screed on membrane completed usually after plastering)

Figure 1.1

MEMBRANES

Building Regulations

Part C of Schedule 1 to the Building Regulations 1985 states that the walls, floors and roof of a building should adequately resist the passage of moisture to the inside of the building.

Regulation C4 states that any ground-supported floor will meet the Regulations if the ground is covered with dense concrete, laid on a hardcore bed, and a damp-proof membrane provided.

The construction of a solid ground floor then may be provided as follows:

(a) Concrete at least 100 mm thick, the leanest permissible mix being 1:3:6 cement:fine aggregate: coarse aggregate.

(b) A hardcore bed of clean broken brick or similar inert material free from materials that could damage the concrete.

(c) A 1000 gauge polythene sheet, or better, laid on top of the hardcore on a bed of sand or similar which will not cause damage to the membrane. This is placed below the concrete.

or

Any suitable membrane such as 1000 gauge polythene or three coats of cold applied bitumen, laid on top of the concrete and protected by a screed or suitable floor finish. Pitchmastic or a similar material may serve as a floor finish and a damp-proof membrane. Where any joints occur in a membrane, these must be well lapped or sealed as necessary. It follows that such a membrane must be continuous with, or joined and sealed to, any barrier to rising damp inserted in any adjoining floor, wall, pier, column or chimney in order to comply with C4.

Note

Where joints occur in a membrane, they must be lapped at least 100 mm, and preferably 150 mm.

A possible source of weakness exists where a floor abuts a wall since, unless the d.p.m. is continuous with the horizontal damp-proof course (d.p.c.) in the walls, a path for rising damp occurs, as shown in figures 1.2 and 1.3. Both situations are quite possible. In each case the d.p.c. is correctly placed, as is the membrane, according to the Building Regulations 1976. The mistake is that there is no continuity between d.p.c. and d.p.m. The situation shown on p. 4, where solid and timber floors are in adjacent rooms, could occur; again, unless the overlap exists between d.p.c. and d.p.m., there is always a possibility of rising damp occurring.

One of the simplest methods of ensuring continuity

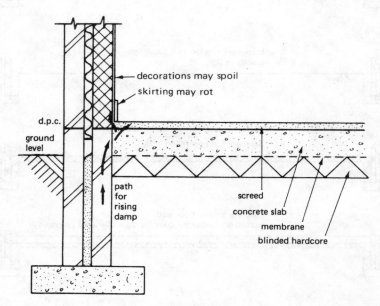

Figure 1.2

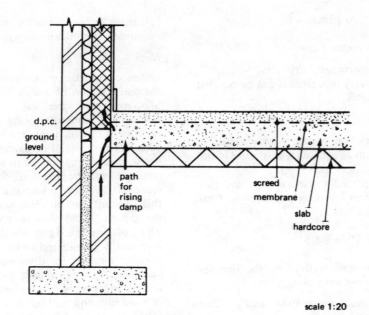

Figure 1.3

scale 1:20

of the two is to build a d.p.c. 225 mm wide into the internal leaf of the cavity wall, and allow this to project 112 mm on the inside. This can easily be turned up or down as necessary to form a lap with the d.p.m. (see figures 1.4, 1.5 and 1.6). Alternatively, the area of wall between the two could be painted with bitumen or coal-tar pitch of sufficient thickness to ensure continuity.

Position of Membrane

The d.p.m. can be placed above or below the slab, each method having its advantages and disadvantages.

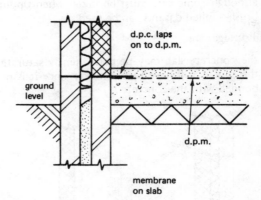

Figure 1.5

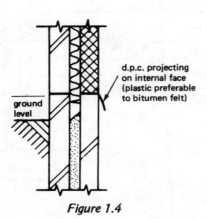

Figure 1.4

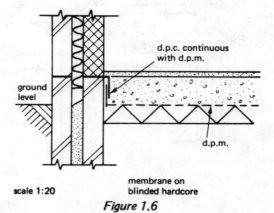

Figure 1.6

Method 1: Below the Slab (figure 1.6)

The advantages of this method are

(1) the slab is kept permanently dry
(2) the screed can be very thin since it can be bonded to the slab. (See table 1.1.)

Disadvantages include

(1) placing the d.p.m. on hardcore may result in punctures unless an adequate layer of sand blinding is present
(2) planks must be used for barrow runs and care must be taken when wheeling and tipping if the d.p.m. is to be kept intact.

Method 2: On the Slab (figure 1.5)

This is the more common method of the two for dwellings and has the following advantages

(1) the membrane is much easier to lay and work on without fear of damage
(2) the concrete slab will dry out much more quickly; it can take up to 6 months on a d.p.m.
(3) planks are not required for wheelbarrows, although some care must be taken when tipping
(4) brush-applied d.p.m.s can be used.

Disadvantages are

(1) the concrete slab may be permanently saturated
(2) the screed will have to be thicker since it is not bonded with the concrete

(3) there is a possibility of failure of the screed unless the thickness is satisfactory. (See table 1.1.)

Note

Unless circumstances are favourable, it is unusual in the construction of a dwelling to place the d.p.m. on blinding below the slab, mainly because of the amount of concrete being placed, the horizontal movement of which entails a great deal of foot traffic, wheelbarrows etc. For example, in an average-sized house measuring, say, 10 m x 10 m, 10 to 15 m^3 of concrete will be required for the floor slab. This represents 200–250 wheelbarrow loads, and moving this amount on blinded hardcore without tearing the d.p.m. would slow down progress considerably. Thus, placing the d.p.m. on the slab below the screed is generally considered to be simpler and is the most popular method for dwellings. For larger buildings, however, such as garages, factories and warehouses, the d.p.m. is more often placed on an adequate layer of sand blinding, and construction joints are carefully made as necessary. Thus the slab, which contains crack-control steel reinforcing, is able to expand and contract without undue stresses occurring, also the steel is kept away from a potentially damp situation where rusting could take place.

ADJOINING SOLID/HOLLOW FLOORS

Floor surfaces within a dwelling should of course be level and continuity of the d.p.c./d.p.m. is important.

Figure 1.7 shows the usual constructional details.

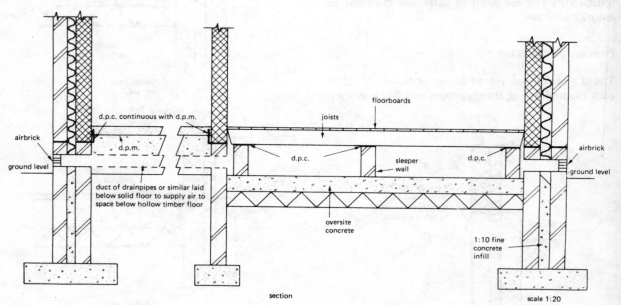

Figure 1.7

Where floors of different construction are incorporated within a dwelling, particular attention must be paid to the provision of a through air supply below the timber floor. This is usually provided by installing a duct of pipes, passing from an airbrick and through any partition walls, before the solid ground-floor slab is poured (see figure 1.7). Care must also be taken to ensure continuity between d.p.c. and d.p.m.

Solid Floors Incorporating Timber

Where it is required to incorporate timber into solid ground floor construction, battens, to which the floorboards can be nailed, must be fastened on or in the screed.

Battens in the Screed

Dovetailed battens are embedded in the screed on the d.p.m. The battens must be either treated in accordance with the provisions of BS 3452: 1962, or impregnated under pressure with an aqueous solution of copper—chrome arsenate, and any surfaces exposed by cutting must be thoroughly treated by dipping, spraying or brushing with an aqueous solution of not less than 10 per cent copper—chrome arsenate (see figure 1.8).

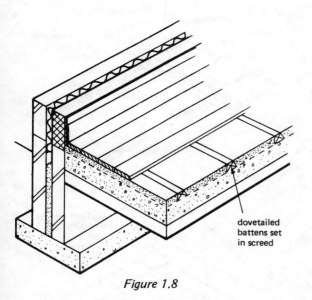

Figure 1.8

Battens on the Screed

Galvanised floor clips must be fixed to the screed without piercing the d.p.m. One method of doing this is to set the clips into the screed, and when this has set and hardened and thoroughly dried out the

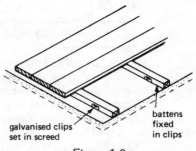

Figure 1.9

battens are laid in these clips and the floorboards can be nailed (see figure 1.9).

FLOOR SCREEDS

A screed can be described as a floor finish or a surface to receive a floor finish depending on the materials used, and it should be perfectly level. Sand and cement are normally used to form a screed, but only where it is intended to be given a finish of tiles or a similar material, since it presents a cold, noisy surface of poor appearance and is liable to dusting. Thus it is not usually considered suitable for a wearing surface in dwellings and if an *in situ* wearing surface is required it should be of concrete, granolithic flooring or terrazzo, etc.

Sand and Cement Screeds

The type and thickness of the intended floor finish to be superimposed determines the mix proportions. 1:3 cement/sand is suitable for thin coverings of linoleum, vinyl tiles, etc.; 1:4 for harder, thicker coverings of clay or quarry tiles; and 1:5 for cork tiles, which can then be pinned to the screed. The water—cement ratio should be kept as low as possible since the inclusion of excess water in the mix results in a poor quality screed with a high shrinkage and weak surface layer; such a mix also proves difficult to lay since 'ponding' occurs during the compacting and trowelling operations.

A simple test for sand and cement used for screeding is to squeeze a sample in the palm of the hand: the sample should hold together and it should not be possible to squeeze out any water. The minimum screed thicknesses are given in table 1.1.

Fully Bonded

Where it is possible to lay the screed within up to 4 hours of completing the base slab a very strong bond is obtained and thus the thickness need only be 20

Table 1.1

Laying conditions	Thickness (mm)
Fully bonded to base	20
Partially bonded to base	30–40
On a damp-proof membrane	50
On a dirty base contaminated with oil or grease	60
On a compressible layer used for sound or thermal insulation	75
On a base containing heating elements	75

mm. The only necessary treatment to the base is to brush the surface to remove any laitance and expose the aggregate. This is known as monolithic construction, and when it is possible to make use of this method there is little possibility of non-adhesion, cracking or curling at the edges. It is unusual, however, for the screed to be laid at this time since, as previously shown (figure 1.1), the slab is poured when the substructure brickwork is completed, whereas the screed is laid after completion of the plastering.

Partially Bonded

When the screed is laid some weeks after the slab, the thickness depends on the equipment available to provide a key. If the base is thoroughly mechanically hacked a thickness of 25 to 30 mm is suitable, but for a hand-hacked surface 40 mm is to be preferred. In either case it is important that the base is perfectly clean, clear of dust and damped before screeding. The use of a bonding agent will help adhesion between slab and screed.

Unbonded Construction

When the slab cannot be hacked, or is covered with grease or a d.p.m., etc., no bond will be obtained between the screed and the base. Thus the thickness must be increased as shown in table 1.1.

Laying Screeds

For large areas the usual practice is to divide up the area into bays of up to about 20 m^2 by bedding accurately levelled battens in the screed material and laying alternate bays in a chequerboard pattern. The bays are preferably kept square and in no case should the length exceed one-and-a-half times the width. On the following day the battens are removed and the rest of the bays can be laid using a straightedge to tamp between the completed bays. To attempt to

complete an over-large area would result in cracking occurring since the slab will shrink slightly as it sets. In dwellings it is usually possible to complete each room in turn and here again, for accuracy, screeding battens should be bedded on the floor and the floor can be screeded between and around the battens (see figure 1.10). Immediately on completion the battens are removed and the lines filled in with screeding material.

If the screed is too thin to use battens a fillet of sand and cement about 300 mm wide can be laid around the room and the floor is screeded to this fillet (see figure 1.11).

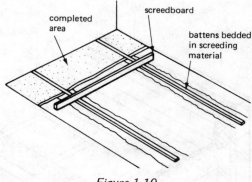

Figure 1.10

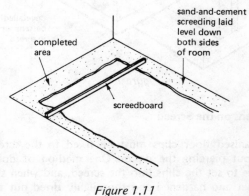

Figure 1.11

Finishes to Screeds

Screeds to receive clay or quarry tiles, for example, can be lightly tamped or finished with a wooden float. Thinner flexible tiles usually require a steel float finish, applied some 2 to 3 hours after the screed has been laid. For this to be accomplished, planks may have to be packed up on bricks strategically placed so that all parts of the floor can be reached. Curing should take place for at least 3, and preferably 7 days after laying.

Causes of Failure

(1) Lack of bond between base and topping: this is a very common cause of failure, especially in separate construction, and when it occurs the topping becomes hollow in parts and will eventually crack and require replacing, either by patching, or taking up the whole floor and relaying, depending on the extent of the failure.

(2) Too large an area laid: if an area larger than that recommended is laid it may crack because of natural drying shrinkage.

(3) Insufficient curing: this may result in excessive shrinkage cracking and a surface liable to dusting. Curing should be carried out by one of the methods explained in Volume 1, chapter 3.

CONCRETE FLOORING

A stronger finish can be achieved with concrete than with cement and sand but once again the finish is unacceptable for dwellings. It is suitable for garages, factories and warehouses and when laid in large areas should contain crack-control steel and contraction joints. The concrete must be laid as dry as possible consistent with workability and is finished with tamping beam and preferably a power float. Here again curing is important and the thickness must be at least 100 mm.

GRANOLITHIC FLOORING

This is laid on a concrete base in a similar manner to cement and sand, the required thickness, depending on the condition of the base, being from 25 to 60 mm. The mix is 1:2½ cement/granite chippings with a little sand if the mix proves too coarse to give a good finish. A very hard-wearing surface can be provided in this way and the wearing properties can be further improved by trowelling carborundum into the surface while it is still wet.

COMPARISON OF HOLLOW TIMBER FLOORS AND SOLID FLOORS

Hollow timber floors and solid floors are illustrated for comparison in figure 1.12.

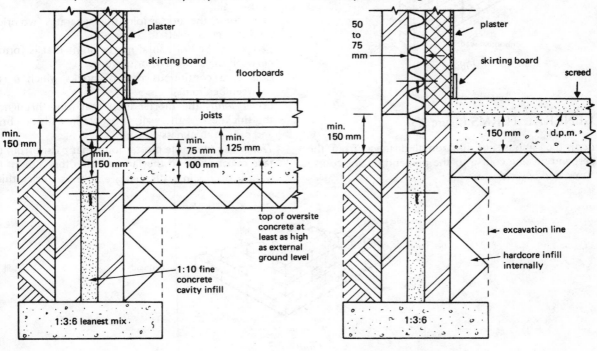

Figure 1.12 Comparison of hollow timber floors and solid floors

2
BONDING BRICKWORK: PATTERNS

To comply with Codes of Practice and the requirements of the Building Regulations when constructing
walls it is essential that the walls are built to have
stability and the capability to accept loads. It is
necessary therefore to understand the reason for
building walls to required patterns.

(1) The term bond when used in brickwork refers to
the arrangement of bricks to a regular pattern.
(2) The reason for bonding walls is to distribute any
intended load throughout the length and thickness of the walling, thereby ensuring both lateral
and vertical stability (see figures 2.1 and 2.2).

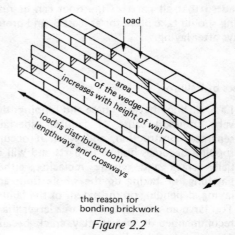

the reason for
bonding brickwork
Figure 2.2

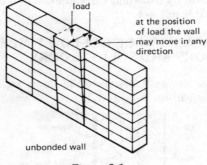

Figure 2.1

GENERAL TERMS (figure 2.3)

Lap, the amount of horizontal distance from the
perpend on one course to the perpend on the course
above and below.

Perpend, a vertical mortar joint appearing on the face
of the wall.
Cross joint, the mortar joint that separates two bricks
on the same course.
Bed joint, the horizontal mortar joint that is formed
between the courses of brickwork.
Course, a continuous row of bricks which occurs
between bed joints.
Collar joint, the mortar joint that runs throughout
the thickness of the wall; it occurs in walls one brick
and over in thickness.
Transverse joint, the mortar joint that passes from the
face on one side to the face on the other side of the
wall; it occurs in walls that are at least 1 brick thick.

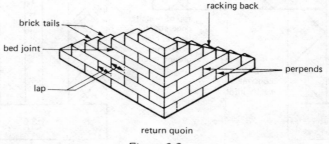

Figure 2.3

Return quoin, the angle formed at the junction of two external walls.

A stopped end, the termination of a wall, which should be correctly bonded to coincide with the bond used for the walling.

TYPES OF BOND

Many types of bond are used in brickwork, but all the varied arrangements are required to have a lap of half brick or quarter brick.

English Bond

This is only used for walls of one brick and over in thickness. The amount of lap is 56 mm or a quarter of a brick, and it is obtained by placing a queen closer next to the quoin header on every alternate course. The bond consists of alternate courses of headers and stretchers; it contains no weaknesses and it is effective both crossways and lengthways. English bond is normally used for walling that requires the maximum amount of strength. The disadvantages with this bond are, however

(1) the unattractive face appearance
(2) courses of alternate headers and stretchers tend

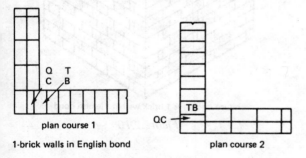

plan course 1

1-brick walls in English bond

plan course 2

Figure 2.4

to have a slowing-down effect on craftsmen so that labour costs are often increased (figures 2.4 to 2.7).

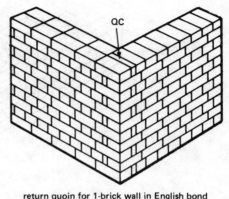

return quoin for 1-brick wall in English bond

Figure 2.6

Flemish Bond

This is another quarter-lap arrangement; the lap is formed by placing the queen closer next to the quoin header on every alternate course. Flemish bond is used for walls of one brick and over in thickness. The bonding arrangement consists of alternate headers and stretchers on the same course. This bond is attractive and may be enhanced by using a variation of colours for either the headers or the stretchers. Although the bond contains internal straight joints which occur at intervals in the wall thickness it is quite suitable for freestanding walls and it possesses considerable lateral strength (figures 2.8 to 2.10).

Stretcher Bond

A very common bonding arrangement which can be used for half-brick walls and also for walls one brick

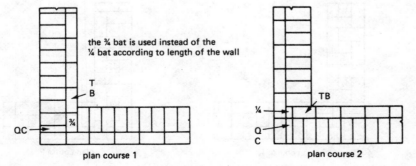

the ¾ bat is used instead of the ¼ bat according to length of the wall

plan course 1

plan course 2

return quoin for 1¼-brick walls in English bond

Figure 2.5

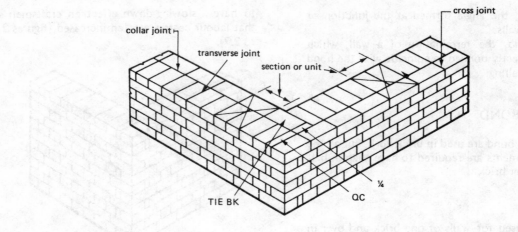

1½-brick English bond quoin showing sectional or unit arrangement

Figure 2.7

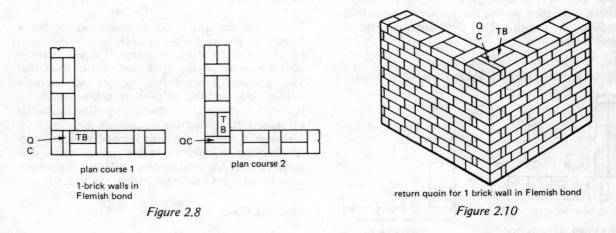

1-brick walls in
Flemish bond

Figure 2.8

return quoin for 1 brick wall in Flemish bond

Figure 2.10

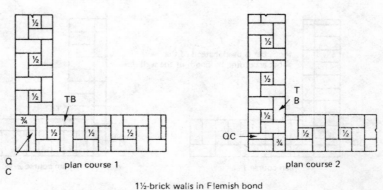

1½-brick walls in Flemish bond

Figure 2.9

in thickness. The amount of lap is half brick, obtained by using half a brick at the stopped end or the quoin header for return quoins. This bond is attractive and usually provides adequate strength for normal forms of construction work; it can also be used with one-third lap (figure 2.11).

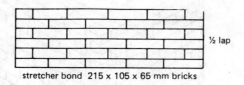

½ lap

stretcher bond 215 x 105 x 65 mm bricks

Figure 2.11

Header Bond

This is normally used for circular brickwork, that is, curved walling; both concave and convex faces can be adequately obtained with the minimum amount of cutting. The amount of lap is quarter brick, achieved by using a three-quarter bat at the stopped end or return quoin, on every alternate course (figure 2.12).

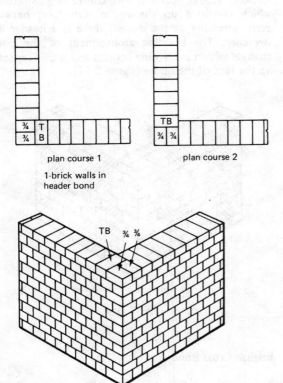

¾ | T
¾ | B

plan course 1

TB
¾ ¾

plan course 2

1-brick walls in header bond

TB ¾ ¾

return quoin in header bond

Figure 2.12

English Garden Wall Bond

A combination of stretcher bond and English bond, the bonding arrangement consisting of three or five courses of stretcher bond with one course of headers; the lap is half brick for the stretcher courses and quarter brick for the heading course, which is obtained by using a queen closer next to the quoin header on the header course (figure 2.13). This bond is often

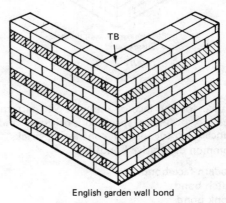

TB

English garden wall bond

Figure 2.13

used for freestanding walls, or walling requiring a decorative face and a reasonable amount of strength, although the internal collar joint is vertical within the height of the stretcher courses.

Flemish Garden Wall Bond

This is commonly termed Sussex bond; the position of the header is after every three stretchers on each course, and the pattern is a header occurring above and below the centre stretcher in each series of three stretchers. There are several bonding arrangements of the quoin, but the strongest with the minimum amount of cutting is shown in figure 2.14. This is a quoin header followed by the queen closer and then three stretchers, header, three stretchers, header, etc.; the alternate course is two stretchers, header, three stretchers, header, etc. The lap is quarter brick; this is a very attractive bonding arrangement, and it can be enhanced in its appearance by the use of contrasting headers.

DECORATIVE FACE BONDS

There are many decorative bonding arrangements, specially designed to provide a pleasing and decorative

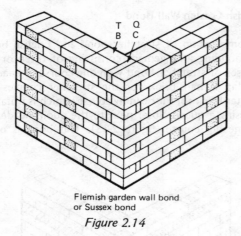

Flemish garden wall bond
or Sussex bond

Figure 2.14

appearance for the face of the wall. Set patterns are arranged to occur in specified positions, and are pronounced by the use of contrasting bricks. Four very common decorative bonds are

(1) modern face bond
(2) Dutch bond
(3) monk bond
(4) English cross bond.

Modern Face Bond

This is a very simple but effective bonding arrangement, using a contrasting header after every two stretchers on each course. The lap is quarter brick, obtained by the use of a three-quarter bat at the quoin on every alternate course (figure 2.15).

Dutch Bond

This bonding arrangement consists of alternate courses of headers and stretchers; a three-quarter bat is used on the quoin on every stretcher course to provide the quarter-brick lap. On every alternate

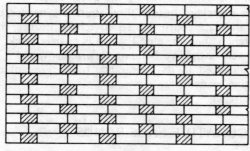

modern face bond

Figure 2.15

stretcher course there is a header next to the quoin three-quarter. Diapers can be formed using contrasting bricks (figure 2.16).

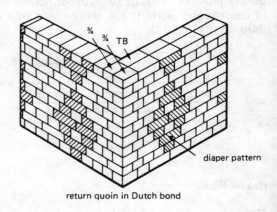

return quoin in Dutch bond

Figure 2.16

Monk Bond

There are variations of this bond, but principally the objective of the bonding arrangement is to provide a vertical zig-zag pattern of contrasting stretchers which continue up the entire wall face; between every stretcher in the diaper, there is a header and stretcher. The bonding arrangement of the quoin changes within every nine courses and is then repeated up the face of the quoin (figure 2.17).

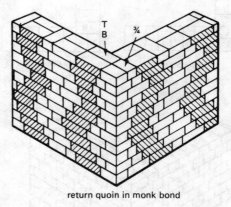

return quoin in monk bond

Figure 2.17

English Cross Bond

This is often referred to as St Andrew's bond. The bond arrangement consists of alternate courses of headers and stretchers, which have a quarter-brick lap. On every alternate stretcher course a header is placed next to the quoin stretcher; a contrasting

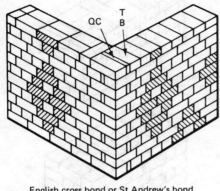

English cross bond or St Andrew's bond

Figure 2.18

diapers formed in stretcher bond

Figure 2.19

repeating pattern of seven courses in height is formed to occur over the entire face of the wall (figure 2.18).

Diapers

This term refers to contrasting patterns formed within the bonding arrangement, which are repeated both vertically and lengthways over the entire wall face; diapers can be formed flush with the wall face, indented or projecting (figure 2.19).

Single Flemish Bond

This is a compound bond, consisting of an external Flemish face and an internal face of English bond. The amount of lap is quarter brick and the minimum thickness of wall is 1½ bricks. The headers on the Flemish face are half or three-quarter bats and the English portion of the wall thickness is 225 mm on the heading course. The bonding arrangement is used when a decorative external face is required and considerable load is to be carried on the inner thickness of the wall (figure 2.20).

Examples of stopped ends are given in figure 2.21.

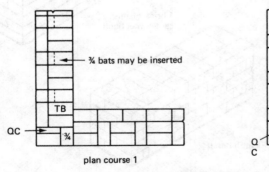

single Flemish bond

Figure 2.20

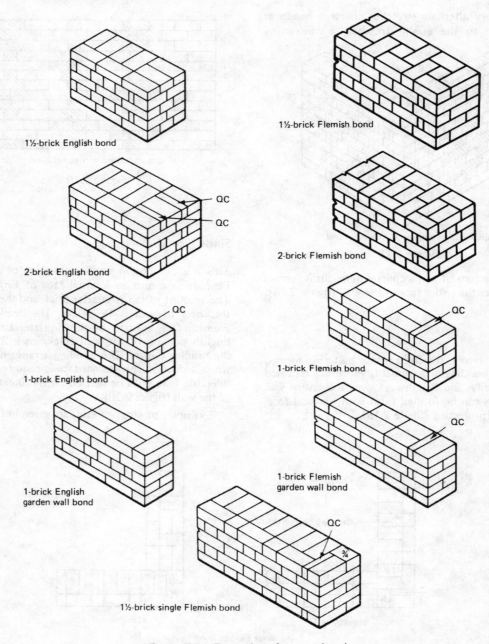

1½-brick English bond

1½-brick Flemish bond

2-brick English bond
QC
QC

2-brick Flemish bond

1-brick English bond
QC

1-brick Flemish bond
QC

1-brick English
garden wall bond

1-brick Flemish
garden wall bond
QC

1½-brick single Flemish bond
QC
¾

Figure 2.21 Examples of stopped ends

3
BONDING BRICKWORK: APPLICATIONS

TEE-JUNCTION WALLS

When bonding tee-junction walls it is necessary to comply with the rules of bonding, that is

English bond: change direction, change face bond; the tie-brick should be a header and the amount of tie should be 56 mm (figure 3.1)

Flemish bond: the tie-brick should always be a stretcher and the amount of tie should be 56 mm (figure 3.2)

For stretcher bond see figure 3.3.

To accommodate the tie-bricks in both English and Flemish bonds it is necessary to insert queen closers within the straight walls at the position of the junction.

INTERSECTING WALLS

When bonding one-brick walls, the position at the centre of the intersection should always contain a header; this permits the required amount of tie to be formed and allows the correct bonding arrangement to be achieved when using both English and Flemish bonds (figures 3.4 to 3.8). The amount of tie is 56 mm for one-brick walls, but can be 56 or 168 mm for walls over this thickness.

RETURN ANGLES

These are illustrated in figures 3.9 and 3.10.

BROKEN BOND

This is the term used to describe the situation when the dimensions of a wall do not work out to actual brick sizes, and therefore a cut brick must be inserted.

Because so many varied situations can occur in the construction of walls it is not possible to provide a rigid format regarding the actual position at which

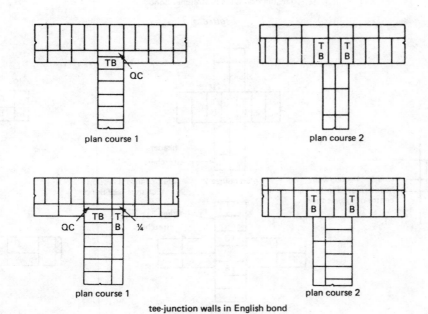

plan course 1

plan course 2

plan course 1

plan course 2

tee-junction walls in English bond

Figure 3.1

15

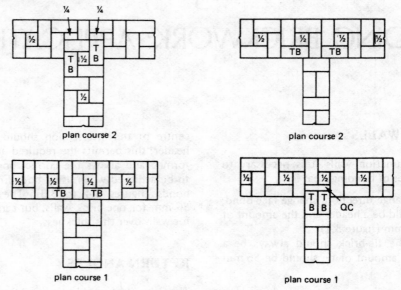

plan course 2

plan course 2

plan course 1

plan course 1

tee-junction walls in Flemish bond

Figure 3.2

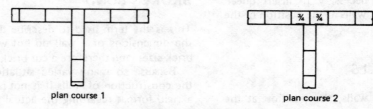

plan course 1

plan course 2

tee-junction walls in stretcher bond

Figure 3.3

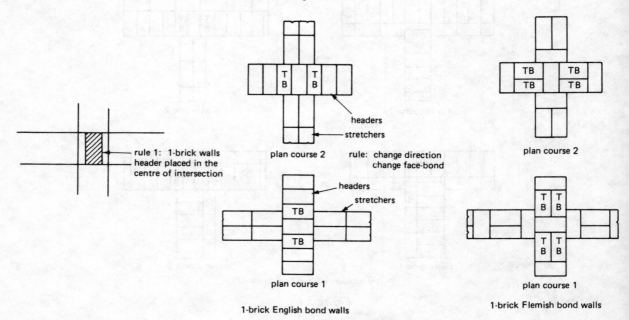

rule 1: 1-brick walls
header placed in the
centre of intersection

headers
stretchers

plan course 2

rule: change direction
change face-bond

plan course 2

headers
stretchers

plan course 1

plan course 1

1-brick English bond walls

1-brick Flemish bond walls

Figure 3.4 Intersecting walls

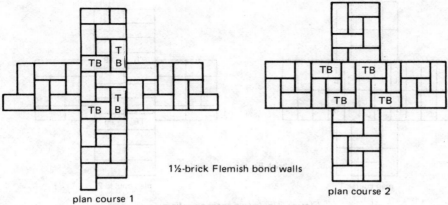

plan course 1

1½-brick Flemish bond walls

plan course 2

Figure 3.5 Intersecting walls

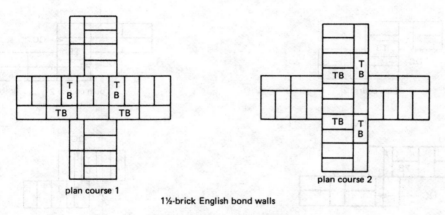

plan course 1

1½-brick English bond walls

plan course 2

Figure 3.6 Intersecting walls

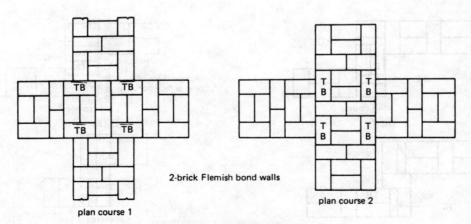

plan course 1

2-brick Flemish bond walls

plan course 2

Figure 3.7 Intersecting walls

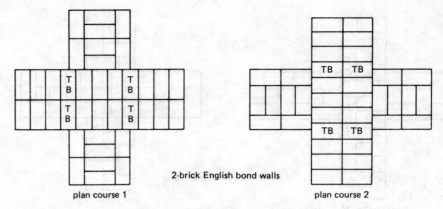

2-brick English bond walls

plan course 1

plan course 2

Figure 3.8 Intersecting walls

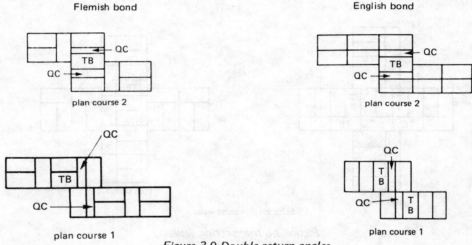

Flemish bond

English bond

plan course 2

plan course 2

plan course 1

plan course 1

Figure 3.9 Double return angles

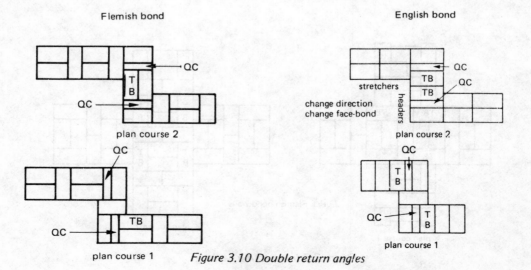

Flemish bond

English bond

plan course 2

plan course 2

stretchers

change direction
change face-bond

headers

plan course 1

plan course 1

Figure 3.10 Double return angles

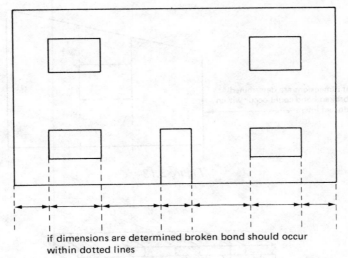

if dimensions are determined broken bond should occur
within dotted lines

Figure 3.11

broken bond should be placed, owing to the following factors

(1) the shape of the wall
(2) the position and dimensions of openings
(3) the bond used for the walling.

It is therefore necessary to provide recommendations which should be complied with whenever the situation allows; these are

(1) broken bond should be placed above and below any opening that occurs within the wall (figure 3.11)
(2) with gables that have raking cuts, the broken bond should occur midway up the angle of rake (figure 3.12)
(3) when walls or gables have raking cuts and contain openings, the broken bond should be placed to occur above and below the openings (figure 3.13)
(4) broken bond should not be placed at quoins, and when walls contain openings and piers that do not work out to brick dimensions, the broken bond should occur above and below the openings and also in the piers (figure 3.11)
(5) whenever possible, broken bond should be formed with cut bricks not smaller than half bats
(6) reverse bond should be considered if it will eliminate the necessity of forming broken bond (figure 3.14).

Further examples are given in figures 3.15 to 3.18.

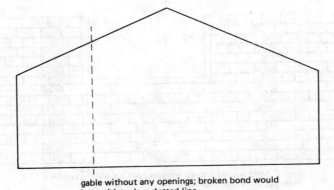

gable without any openings; broken bond would
be positioned on dotted line

Figure 3.12

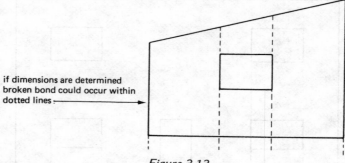

if dimensions are determined broken bond could occur within dotted lines ⟶

Figure 3.13

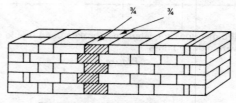

Flemish bond wall with ¾ bats forming broken bond, using reverse bond

Figure 3.14

stretcher bond wall with fixed door opening; walling contains broken bond

Figure 3.15

English bond wall with fixed window opening; walling contains three positions of broken bond

Figure 3.16

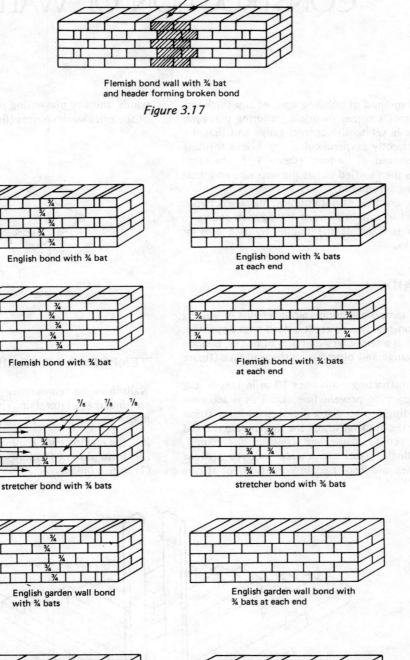

Flemish bond wall with ¾ bat
and header forming broken bond

Figure 3.17

English bond with ¾ bat

English bond with ¾ bats
at each end

Flemish bond with ¾ bat

Flemish bond with ¾ bats
at each end

stretcher bond with ¾ bats

stretcher bond with ¾ bats

English garden wall bond
with ¾ bats

English garden wall bond with
¾ bats at each end

Flemish garden wall bond with
¾ bats

Flemish garden wall bond
with ¾ bats

Figure 3.18 Examples of broken bond between openings when dimensions are fixed

4

CONSTRUCTION OF WALLS

The correct method of building walls of any thickness is first to erect a corner or quoin, ensuring that each quoin brick is set to the correct gauge and that the quoin is perfectly perpendicular. The bricks forming the quoin should all be level (figure 4.1). The same operation is then carried out at the opposite end, that is, to the length required.

After the erection of both quoins, the area between is then filled in with brickwork, laying each course to line, which is fixed at the quoins by line pins or corner blocks.

USE OF AIDS

To provide temporary quoins along lengths of walling it is common practice to erect a dead man or profile. A dead man is a temporary erection of bricks, bedded to correct gauge and plumb with the wall face (figure 4.3).

When constructing walls over 10 m in length it is often necessary to prevent line sag. This is achieved by fixing a tingle, placed at a pre-determined position on the wall; the brick on which the tingle plate is fixed should be set to gauge and checked for plumb. Although tingle plates are mainly used to prevent line sag, they are also used in conditions of strong winds, thereby preventing any distortion in alignment of the brickwork courses (figures 4.2 and 4.4).

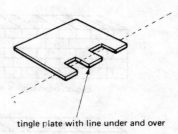

tingle plate with line under and over

Figure 4.2

TEMPORARY TERMINATION OF WALLS

Walls may be temporarily terminated in length and continued at a later date. The three common methods in use are

(1) the racking back method
(2) formation of toothings
(3) block indents

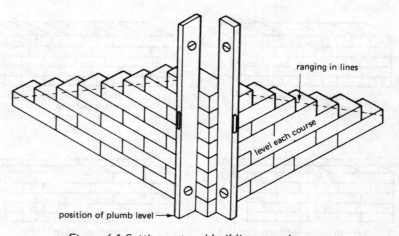

ranging in lines

level each course

position of plumb level

Figure 4.1 Setting out and building a quoin

22

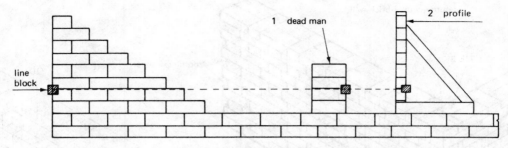

Figure 4.3 Extending quoins using temporary aids (alternative methods shown)

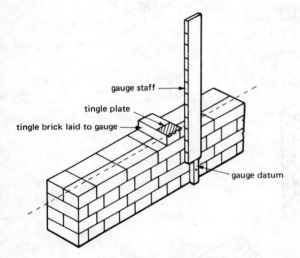

Figure 4.4 Preventing line sag

method is that the height to which it can be effectively used is limited (figures 4.1 and 4.3).

Toothing Method

This method is very often used by craftsmen; it allows for a wall to be temporarily terminated, and can be used for heights not exceeding 1.0 m. When joining up to the existing work at a later date considerable care should be taken to prevent unequal settlement occurring. Toothing is not recommended for loadbearing brickwork, and permission must be obtained from the engineer in charge before it can be used (figures 4.5 and 4.6).

Racking Back Method

This is commonly used; it allows work to be joined up at a later date and provides no problems when the joining up is required. The disadvantage with this

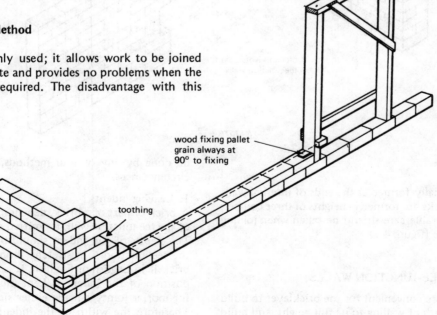

Figure 4.5 Using door casing as a profile or dead man

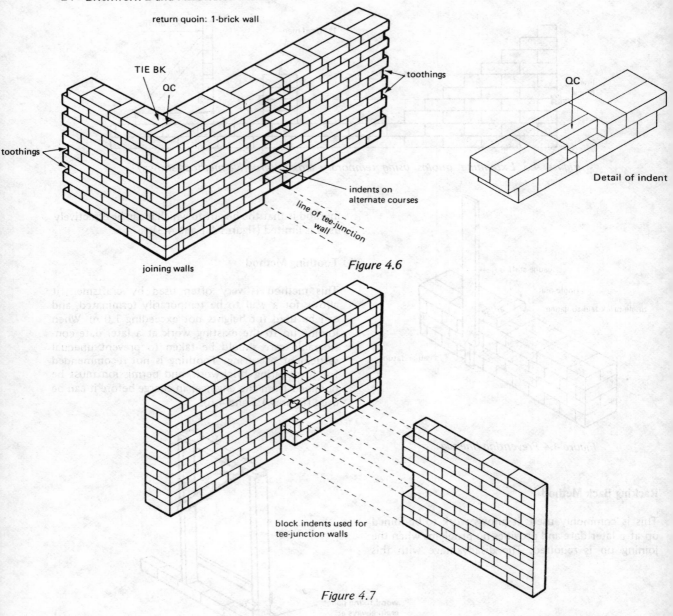

return quoin: 1-brick wall

TIE BK

QC

toothings

toothings

QC

Detail of indent

indents on
alternate courses

line of tee-junction
wall

joining walls

Figure 4.6

block indents used for
tee-junction walls

Figure 4.7

Block Indents

These are usually formed at the ends of internal walls
only; the blocks are formed in heights of three courses;
again considerable care should be taken when joining
up takes place (figure 4.7).

FORMING TEE-JUNCTION WALLS

It is often more convenient for the bricklayer to build
a straight length of walling to its full height, and build
any adjoining walls at a later date. Provision for tying
in the tee-junction wall must be provided and this can

be done by any of four methods, depending on the
circumstances.

1. Leaving indents

A brick or part of a brick, depending on the thickness
of the tee-junction wall, is omitted every other course.
The depth of the indent should be one half brick if
the tee-junction is in stretcher bond, but one quarter
brick if in English or Flemish. It is important that the
position of the indent is kept plumb, and provision
for mortar joints is made either side of the tie brick.
Therefore the width of the indent must be equal to
the thickness of the wall to be built, plus two mortar
joints (figure 4.6).

2. Leaving block indents

These are not so-called because a block wall is to be built, but because of their size. It is often considered quicker to build three courses solid, then leave a 3-course indent. A brick is usually stood on end in the indent while building the course above, and this should be removed at the end of the day before the mortar has set hard. As with indents, block indents must be carefully measured out and kept perfectly plumb, or cutting out will be necessary at a later date (figure 4.7).

3. Leaving a chase

This is not as common a method as the first two. It is more often used where tee-junction walls are built against existing walls. Two vertical lines are marked on the wall to be chased; these are cut to the required depth using an angle grinder, and the centre portion is rapidly disposed with using a lump hammer and bolster. Continuous chases are more often associated with sinking pipework or electrical cables into a wall, as they do not have the strength of indents or block indents (figure 4.8).

4. Building in expanded metal ties

Metal ties of bricktor or exmet, for example, are built into the main wall every six courses. These should be approximately 400 mm in length and are left projecting from the wall face. Using this method the bonding pattern, whether of brickwork or block-work, can be carried straight through and no cutting is involved. As with indents, the expanded metal ties must be carefully positioned to coincide with the partition wall, and successive ties must be kept plumb above the first (figure 4.9).

TYING TO EXISTING WALLS

Where a wall is to be tied at right angles to an existing wall, the modern method is to use wall extension profiles. It is not only quicker than cutting out, but is much cleaner and causes far less disturbance. Double-flanged profiles of galvanised steel (internal walls) or single-flanged profiles of stainless steel (external walls) are clamped to the existing wall into prepared plugs (figure 4.10). The tee-junction wall is built into these profiles, inserting the special wire wall ties provided into the fixing slots every 300 mm max. (figure 4.11). The 'Furfix' wall extension system is a most convenient, labour-saving device for use in this situation, and may be used to provide lateral support for tee-junction walls up to three storeys in height (8 m max.).

Fixing procedure is as follows:

1. The profile is offered up to the existing wall with the wall ties aligned with the proposed bed joints.

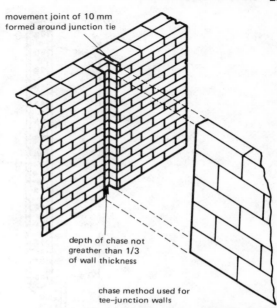

movement joint of 10 mm formed around junction tie

depth of chase not greather than 1/3 of wall thickness

chase method used for tee–junction walls

Figure 4.8 Tying in junction walls

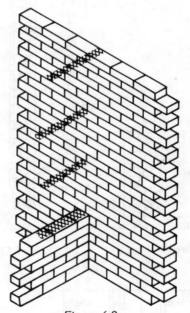

Figure 4.9

The profile is plumbed and fixing holes are marked as shown in figure 4.10.

2. The existing wall is drilled and plugged (into bricks, not joints) and the profile is clamped to the wall with the coach screws provided.

3. Full cross joints are placed on the bricks or blocks before bedding in the normal manner against the profile (figure 4.12). Wall ties are positioned in the profiles and embedded into the bed joints at 300 mm max. centres.

BRICKLAYING

Accuracy in Building

To ensure continuity of accuracy during the construction of a building, it is considered good practice to carry out a setting out and checking procedure at ground level. This requires all walls to be checked for level—normally using the Cowley level—and all dimensions to be checked using a steel tape adequately supported throughout its length. All angles are checked for squareness; this can be done using a large builder's square but, if the angle governs a line over 15 m in length, it should be checked with the use of an instrument. It is essential to construct brickwork with care and accuracy and to comply with the recommended tolerances shown below

Length	
up to 5.0 m maximum	∓ 5 mm
between 5 and 10 m	∓ 10 mm
over 10 m	∓ 15 mm
Height	
up to 3.0 m maximum	∓ 5 mm
between 3 and 6 m	∓ 10 mm
over 6 m	∓ 15 mm

Uniformity

During building operations it is important to ensure uniformity of settlement. This is achieved by building all walls in a uniform manner and ensuring that no wall or section of the building rises more than 1.0 m above the level of the other walling. The maximum height to which brickwork should be erected is not

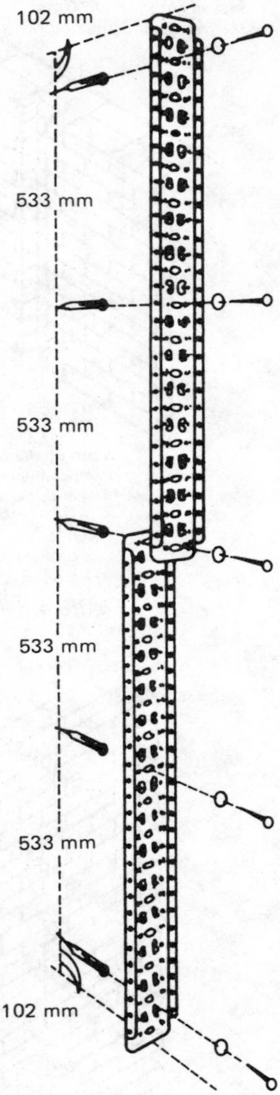

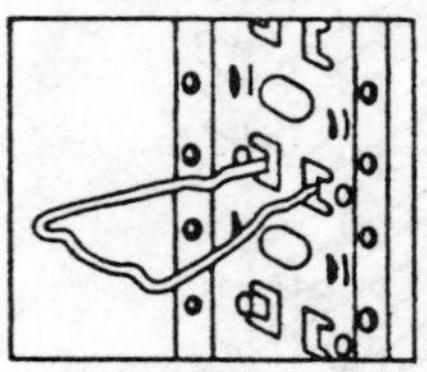

wall tie detail
ties at 300 mm centres

Figure 4.10

102 mm
533 mm
533 mm
533 mm
533 mm
102 mm

Figure 4.11

Figure 4.12

eye line along bed joints
checking perpend
inspection of brickwork

Figure 4.13

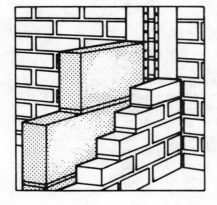

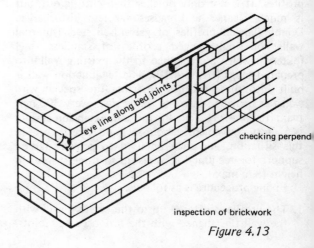

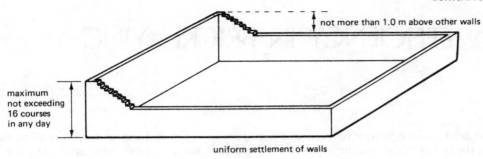

Figure 4.14

exceeding sixteen courses in one day, but this can be exceeded with the permission of the structural engineer (figure 4.14).

For loadbearing brickwork formed with cavity walls, the settings should not exceed 450 mm, that is, each wall should never rise more than 450 mm above the other leaf or wall at any time (figure 4.15).

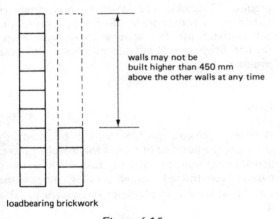

Figure 4.15

Laying Bricks

Bricks that contain a single frog should be laid with the frog uppermost and on a full mortar bed. When bricks have a double frog the shallow frog should first be filled with mortar and laid downwards.

No cross joints, collar joints or transverse joints should exceed 13 mm in thickness.

Before bricks are laid, the engineer may wish to adjust the suction rate of the bricks. This is achieved by wetting the bricks using spray or jets; under normal conditions 2 kg/m^2 is considered a reasonable minimum but this may be altered according to the condition of the bricks, and the weather at the time of the bricklaying operations. Cutting of bricks is normally carried out using the bolster chisel and lump hammer, although the brick hammer and brick trowel are often used for rough work. When perforated or cellular bricks are used, accurate cutting can only be achieved using a masonry saw. Perforated and cellular bricks are always laid on a full mortar bed and the perforations are not filled unless required by the engineer in charge of the project. At the termination of work it is essential to protect the top courses of brickwork from bad weather and to prevent frogs and perforations from being filled with rain. Therefore frogs may be laid downwards and perforations filled on the topmost course only.

When non-loadbearing walls are used in conjunction with loadbearing walls and both are constructed at the same time, the top of the non-loadbearing walls should be finished with a compressible layer of at least 10 mm between the top of the wall and the floor or beam that is to be supported (see chapter 12, on blockwork).

Working Overhand

To obtain maximum productivity from the labour force available, that is, bricklayers and labourers, it is essential to place bricklaying materials on one side of the wall only. At ground level the method of working is often determined by site conditions, for example, if the external ground level can be made suitable. This allows the materials to be placed on the external face of the walling, and the bricklayers erect the internal walls first to a height of approximately 1.3 m. The external wall is then built up to the level of the inner leaf. When this method is not practicable, the work is carried out by loading the internal ground floor with materials. The bricklayers then work from the inside of the building and erect the external walls first. The same method of working can be applied to situations when scaffolding is used to provide the working platform.

Whatever methods are used, it is essential to prevent fatigue for the craftsmen, therefore materials should have a continuous flow and should be placed at a convenient distance from the working area.

5
EFFICIENCY IN BRICKLAYING

Recognising that craft training is now reduced to 3 years and that at the end of this period the apprentice is expected and often required to work with craftsmen of many years' experience, also, that many students of building expect to aspire to positions of site management, the authors emphasise the importance of efficiency in craft operations and the necessity to use 'method building' whenever the opportunity is available.

Efficiency will obviously improve productivity, but it can only be obtained if the following can be achieved

(1) all craftsmen on site should accept 'method building'
(2) standardised craft methods of erection should be accepted by all craftsmen on site
(3) building techniques should be standardised and used whenever possible
(4) all craftsmen should be aware of the plan of building operations, and accept the involvement that is required.

PRODUCTIVITY

Methods of Erection

The building of quoins by bricklayers often varies, and is not consistent even when situations are identical. The building of quoins requires the bricklayer to use the plumb level, gauge staff and tools on almost all bricks used in the quoin; each brick and course receives special treatment, therefore the bricks used in the quoin can be termed 'costly bricks'. It is estimated that 15 per cent of the bricklayer's time is spent in plumbing and levelling, and the higher a quoin is built the more time is spent by the bricklayer in operations which can be considered to be unproductive. Figure 5.1 shows a wall erected with a large setting and two small settings, the shaded areas indicating the bricks that have received special treatment, and which are 'costly bricks'. The large setting requires 136 bricks, the two small settings require 72 bricks. The latter method produces a saving of approximately 48—50 per cent, which substantiates the fact that small settings, built to regular heights, will provide efficiency and greater productivity.

Profiles

Whenever possible profiles should be used; they eliminate the building of quoins and therefore increase actual laying time. Profiles may take the form of door casings, constructed timber profiles, purpose-made metal profiles or even planks or struts (figures 4.3 and 4.5).

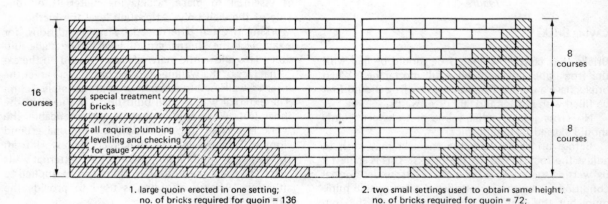

16 courses

special treatment bricks

all require plumbing levelling and checking for gauge

1. large quoin erected in one setting;
   no. of bricks required for quoin = 136

8 courses

8 courses

2. two small settings used to obtain same height;
   no. of bricks required for quoin = 72;
   total saving = 64 bricks

Figure 5.1

Employment of Bricklayers

Often on many building sites craftsmen are employed building walls that are 'too long' for the number of craftsmen working on them. This practice results in unnecessary fatigue and reduced productivity; conversely, situations occur where too many craftsmen are engaged on walls that are not of sufficient length, resulting in men being under-employed. Careful planning is therefore required in the distribution of craftsmen and the area of work involved. Figures 5.2 to 5.4 show examples of the economical use of bricklayers.

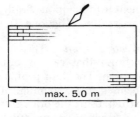

max. 5.0 m

1 bricklayer: economical
2 bricklayers: uneconomical

Figure 5.2

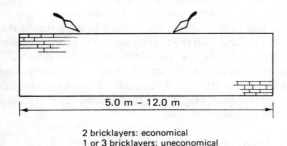

5.0 m – 12.0 m

2 bricklayers: economical
1 or 3 bricklayers: uneconomical

Figure 5.3

Methods of Working

Figure 5.4 shows two bricklayers working on one wall: one starts work at one end, and works towards the centre, while the other man works from the centre to the end; the men do not come into each other's working area, consequently tool actions and laying techniques are normal and unhindered. Figure 5.5

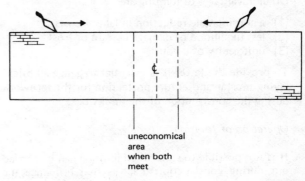

¢

uneconomical
area
when both
meet

Figure 5.5

shows both men working from the ends and meeting in the centre of the wall. This inevitably leads to a restriction in tool actions, and slows down the laying rate; the centre of the wall becomes a low productivity area, showing that some form of planning is required during the building of simple straight walls.

JOINTING AND POINTING BRICKWORK

The above terms refer to methods of finishing the mortar joints that occur on the face of a wall.

Jointing

This term is used to describe the operation of providing the finish to the mortar joints during the con-

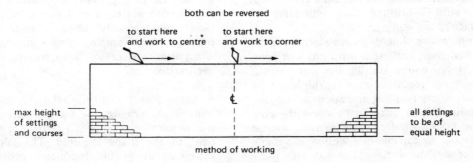

both can be reversed

to start here
and work to centre

to start here
and work to corner

¢

max height
of settings
and courses

all settings
to be of
equal height

method of working

Figure 5.4

struction of the walling. The joint finish is applied using the same mortar that is used for bedding the bricks. The jointing operation is carried out during or after laying each course of bricks, or when the mortar joints are sufficiently hard to receive a finish. Jointing is always preferred to pointing for load-bearing brickwork, or when mortar strength is essential.

Other advantages of jointing are

(1) a considerable reduction in labour costs
(2) less likelihood of damage by rain or frost
(3) uniformity of colour.

To provide the latter it is essential to gauge all brick-laying mortar and ensure protection for the facework during the construction of the brickwork.

Operation of Jointing

It is not possible to state when jointing can be carried out during construction. Factors that influence the operation are

(1) types of brick being used
(2) suction rate of the bricks
(3) type of mortar used for the walling
(4) weather conditions at the time of erection
(5) type of bond used for the walling
(6) thickness of the wall.

Pointing

This term refers to the finishing of mortar joints after the structure has been erected, or when a structure needs to be re-pointed after considerable deterioration of the mortar joints.

To carry out this operation it is essential to ensure that all joints are raked out to a depth of not less than 12 mm and not exceeding 19 mm. Special joint scrapers are used to rake out newly laid mortar and this operation should only be carried out after at least two further courses of brickwork have been laid. After complete raking out has been completed, the walling should be brushed down with care, not to stain the face of the walling.

Pointing operations should always start from the top of the building, working downwards and completing each storey before starting the work below.

Walls sometimes require cleaning. This must be carried out before any pointing operations are begun.

Before pointing it is essential to assess the condition of the brickwork face. The existing mortar and bricks should be capable of adhesion and therefore it is necessary to dampen the wall face using spraying techniques; this provides a uniformity of

damping and prevents over-wetting, which would result in shrinkage of the newly formed mortar joints.

Types of Pointing Joint

From the joint profile (figure 5.6) it will be seen that there are ten recognised finishes to mortar joints. The types most commonly used are 1, 2, 3, and 4.

Since brickwork courses are laid to line it is generally accepted that if there is only the slightest variation in the thickness of the bricks it will be seen to occur along the bottom arris of each course, therefore whatever degree of skill is exercised by the craftsman, it will be exceedingly difficult to obtain perfect alignment of the joint finish when certain types of joint are required (figure 4.9).

1. External Weather Joint   This range is the best type of joint finish that can be obtained to resist weather penetration. The joint profile shows the top of the joint inclined to a maximum of 2 mm and the finish can be struck using the brick trowel or pointing trowel. Surface hardness is assured because of the compaction obtained from the pressure exerted by the trowel.

2. Internal Joint   This joint is completely opposite to the weather joint, since the inclination is formed at the bottom of the joint, to a maximum of 2 mm. This joint is extremely attractive because the top of the brickwork courses are aligned and the top of the joint is flush with the bottom arris of the course above, which conceals any variations. It is extensively used when fair-faced brickwork is required internally. The joint is not often used for external work because the ledge formed along the bottom of the joint provides a resting place for water.

3. Flush and Rubbed Joint   This type of finish is achieved by leaving the joint flush from the trowel as the bricks are laid, then filling in any voids with a rubber or sacking pad using a light circular motion. The rubbing action often brings particles of sand to the surface of the joint, therefore it is not recommended for positions of severe exposure. It can be used advantageously when a rusticated finish is required, for example, using hand-made bricks, or when the wall surface is to receive some form of decoration.

4. Keyed Joint   This is a relatively modern type of joint finish, achieved by using a purpose-made half-round jointing tool or a 12–19 mm mild-steel rod bent to the shape required. The joint section, as shown on the profile, is concave, providing concealment for any form of distortion in course alignment.

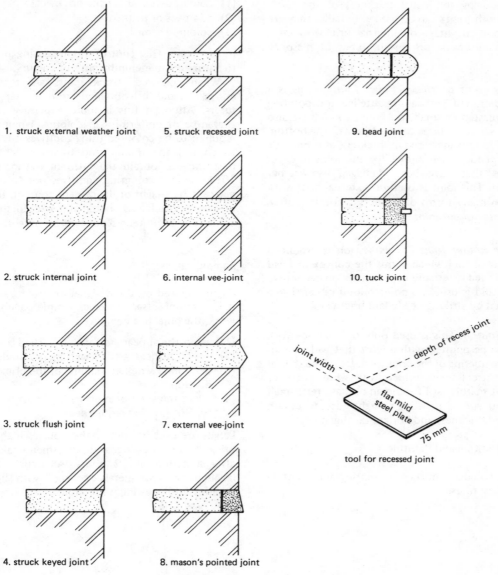

1. struck external weather joint
5. struck recessed joint
9. bead joint

2. struck internal joint
6. internal vee-joint
10. tuck joint

3. struck flush joint
7. external vee-joint

joint width
depth of recess joint
flat mild steel plate
75 mm
tool for recessed joint

4. struck keyed joint
8. mason's pointed joint

Figure 5.6 Jointing and pointing profiles

This finish is now extensively used for both external and internal work; because of the amount of compaction obtained when striking the joint, it is capable of resisting the penetration of weather.

5. Recessed Joint   This joint finish can only be used to advantage on walls built with bricks containing the minimum variation in thickness and is not recommended for loadbearing brickwork or for walls in positions of severe exposure. The recess is formed using a purpose-made steel tool, which scrapes the

mortar from the front of the joint. The depth of the joint should not exceed 19 mm. Forming the joint with a shaped piece of wood is not good practice; the wood becomes worn and the correct depth and width of joint soon decrease.

6 and 7. Internal and External Vee-Joints   The joints are formed using purpose-made jointing tools to provide the type of vee required. The advantage of both joints is in reducing the thickness of the mortar joint; the centre of the vee creates a third or centre line, thereby reducing the visual thickness of the joint. Although both joints can be used externally, they are used to good advantage on internal brickwork of a decorative nature or on certain types of masonry work.

8. Mason's Joint or Heavy Relief Joint   This is a weather joint, cut with a straightedge and pointing trowel or plasterer's small tool; the perpends are also cut to the same width as the bed joints. The bottom of the bed joint is increased in thickness as is one side of the perpend, which allows for the joint to cover any arrises that have been damaged by weather penetration. This joint is used on external brickwork and also on many forms of masonry work. It is often called a *struck* or *cut joint*.

9. Bead or Round Joint   This section is produced using a special tool which forms the convex or bead finish. It is used externally or internally on decorative brickwork and is usually a pointed joint rather than a joint formed by striking newly laid brickwork.

10. Tuck Joint   This is used only on old brickwork which is to be pointed with a joint that will camouflage the weathering of the brickwork face. The special tuck or centre line is formed with white putty or ground soft marble. A Frenchman or plasterer's small tool is used with a pointing straight edge, to obtain the required thickness and horizontal alignment.

Coloured Jointing and Pointing

To obtain coloured mortar for jointing and pointing it is necessary to use

(1) coloured mortar for the brickwork
(2) additives or pigments
(3) coloured cements.

In each case strict control and measuring are required to ensure complete uniformity of colours.

When coloured struck joints are required the mortar should be obtained from mortar manufacturers. Although this is more expensive, there is a guarantee of uniformity in colours. When mortar is required to be coloured using additives it is necessary to control the mixing and to use strict forms of measuring to obtain the ratio of the materials. BS 1014 requires colouring additives not to exceed 10 per cent by weight of the cement content. In the case of carbon black this should not exceed the cement content by more than 3 per cent by weight.

Sample Panels

When required by the designer or structural engineer, it is often necessary to erect sample panels of brickwork, the objective being to

(1) assess the appearance of the finished brickwork
(2) establish the degree of wetting required
(3) determine whether staining of the brickwork will develop
(4) select the type of mortar, thickness of joint and type of jointing required.

Panels are usually built twelve courses high and four bricks wide, viewing and assessment taking place after a period of 72 hours. All panels should be labelled and completely covered with lightweight sheeting until viewing is required.

6

CAVITY WALLING

In modern building construction it is a requirement of the Building Regulations 1985 that all external walls of dwellings are built in cavity form, that is, two skins or leaves of brickwork are built with a cavity between the leaves. The width of the cavity may be a minimum of 50 mm or a maximum of 75 mm (more in certain cases—see Building Regulations), with a minimum overall of 250 mm.

The advantages of this form of construction are

(1) a dry interior is assured
(2) the enclosed air space in the cavity provides the interior of the building with an equable temperature
(3) the inner leaf may be constructed with a different material, which may increase the thermal efficiency and may reduce cost in labour and materials
(4) since cavity walling is usually built in stretcher bond, productivity is therefore increased.

Obviously there are slight disadvantages with cavity wall construction; these are

(1) special treatment is required around all door and window openings
(2) sealing of the cavity must occur at roof level, which may increase labour costs.

WALL TIES

All cavity walls are required by the Building Regula-

tions to be stabilised. This is achieved by the use of galvanised, non-ferrous or plastic ties, which must be fixed into the mortar bed joints of the brickwork during erection.

Types of Wall Tie (figure 6.1)

There are two common types of heavy tie: both have fish-tail ends and are suitable for loadbearing brickwork or bricks containing frogs; they are

(1) fish-tail cranked pattern
(2) fish-tail with single or double twist

Other types of wall tie in common use are

the butterfly tie (figure 6.1)
the Kavi tie (figure 6.1)

The butterfly tie is the most common type of tie in use; it is the cheapest and also provides the required strength for most situations. The tie is formed with 3 mm galvanised wire and is therefore easily accommodated even in mortar beds of thickness 6 mm.

The Kavi tie is made of polypropylene, with copper wire within the plastic. It can easily be accommodated within the thickness of the mortar bed, but it is considerably more expensive than the butterfly tie (figure 6.1).

Position of Wall Ties

The Building Regulations require wall ties to be inserted in all cavity walls at specified positions; these

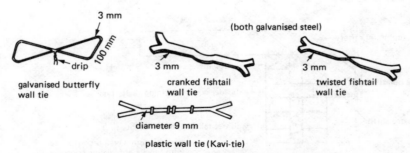

Figure 6.1 Wall ties for cavity walls

33

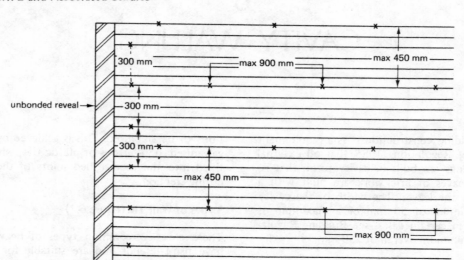

Figure 6.2 Wall ties for cavity walls

are 900 mm apart horizontally and 450 mm vertically. When used at the side of an unbonded reveal they should not exceed 300 mm vertically. It is considered good practice to stagger the position of the ties, thereby ensuring stability for the entire wall area (figure 6.2).

Wall ties should continue up to the terminal point of the cavity wall and be placed if necessary below the tie-brick course at eaves level.

TERMINATION AT EAVES LEVEL

Cavity walls are terminated at eaves level with a sealing course of headers or blocks laid flat; this is used for

(1) sealing the cavity (figure 6.3)

(2) distributing the load of the roof on to both the internal and external walls.

The wall plate is placed directly on the sealing course or it can be placed on one or two courses above the sealing course; the position is determined by the pitch of the roof.

CLEANING CAVITIES

The normal method of preventing mortar droppings from falling to the base of the cavity wall is to use a cavity batten or lath. This batten is placed on the wall ties while one leaf is built to a higher level; the ends of the batten are attached to cords or flexible wire, allowing the batten to be lifted and the mortar droppings to be removed (figures 6.4). The batten should

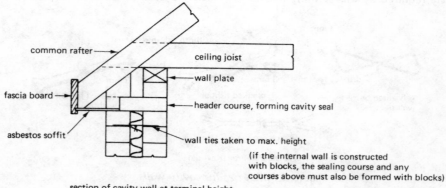

section of cavity wall at terminal height

Figure 6.3

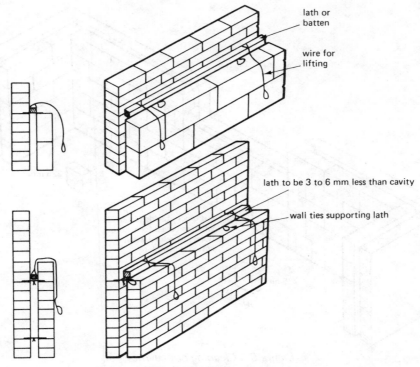

lath or batten

wire for lifting

lath to be 3 to 6 mm less than cavity

wall ties supporting lath

Figure 6.4 Method of preventing mortar droppings

have a tolerance of between 3 and 6 mm less than the width of the cavity.

To provide additional safeguards, coring holes can be formed at intervals along the base of the wall. This allows for easy removal of droppings, which may accumulate at ground level. It is essential that any mortar droppings are removed at the termination of each day's work. The coring hole can be formed using a Welsh arch or sand courses (figures 6.5–6.7). When the brickwork of the structure has been completed the coring holes should then be made good.

THRESHOLDS

The threshold is the position formed at the bottom of a door opening. It is subjected to abrasive wear and must resist the penetration of moisture, while being of good appearance. The materials in common use are terracotta, dense pressed bricks, concrete, stone or quarry tiles.

Thresholds should always be formed using cement mortar and the level of the completed threshold should be at the finish of ground-floor level. Damp-proof courses should be placed under thresholds, ensuring that dampness cannot penetrate into the interior of the structure. The d.p.c. should extend at least 150 mm on either side of the opening (see figure 6.8).

WINDOW OPENINGS AND WINDOW FRAMES

Whenever window openings occur in a brick wall the bonding of the reveal should always be treated as a stopped end. Window frames can be of timber, metal or a combination of both.

When possible, it is considered good practice to build in frames as the work proceeds. This provides economy in labour for the craftsman, for example, a reduced amount of plumbing is required at the reveal; improved fixing techniques and tolerances are not required when forming the opening.

Glazing and temporary covering of the opening can be carried out much sooner, thereby preventing the penetration of water and draughts and providing an equable temperature for internal work to be carried out.

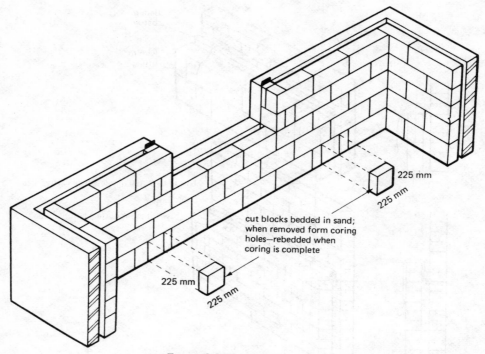

Figure 6.5 Cleaning out cavities

cut blocks bedded in sand; when removed form coring holes—rebedded when coring is complete

225 mm
225 mm
225 mm
225 mm

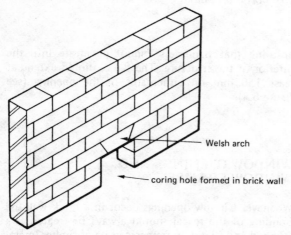

Welsh arch

coring hole formed in brick wall

Figure 6.6

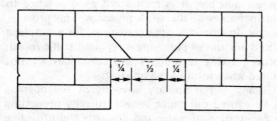

¼   ½   ¼

detail of Welsh arch

Figure 6.7

Fixing Wooden Frames

The joggles on the frame, both top and bottom on both sides, should be first rebated, then knotted, and priming of the frame should then be carried out. The frame is then bedded in mortar, positioned, plumbed and checked for alignment. If a structural sill is used, the sub-sill on the window frame should be fitted with a water bar to prevent capillary action. As the brickwork proceeds, timber pallets are built into the reveal at every 450 mm vertically (see figure 6.9).

It is essential that the brickwork is completely hard before nailing of the frame is carried out. Care should be taken to ensure that load is not imposed on the head of the window frame.

Fixing Metal Frames

It is possible with this type of frame to carry out the fixing at a later date, that is, when the structure has been completed, but obviously better security of fixing can be achieved if the fixing is carried out as the work proceeds. Fixing is carried out by means of metal lugs, which are secured and bolted into slots formed in the side of the metal stile. The slots in the stile allow for alignment of the lug to work in with course levels; tightening the screws is done only to finger tightness. When frames are fitted after com-

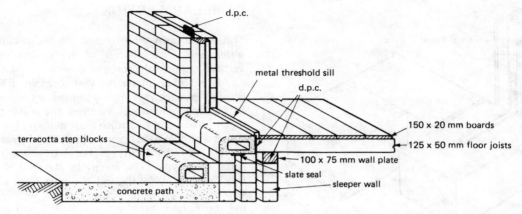

Figure 6.8 Threshold construction

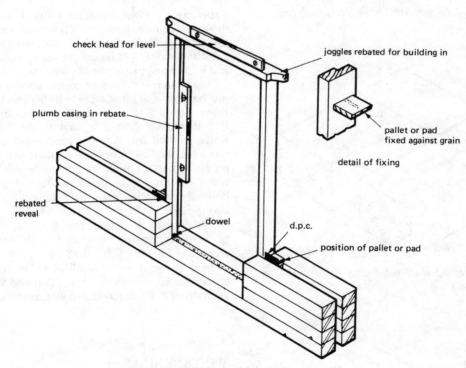

Figure 6.9 Method of fixing door casing

pletion a tolerance of 2 mm per reveal is normally provided (figure 6.10–6.12).

After checking for plumb and alignment it is advisable to check any opening lights for ease of movement. The operation of fixing is complete when a mastic joint is formed between the metal frame and the reveal.

Figure 6.13 illustrates a timber door frame built in with a galvanised metal strap.

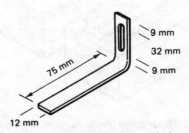

Figure 6.10 Metal adjustable lug

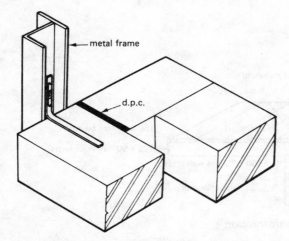

Figure 6.11 Metal window frame fixed with lug

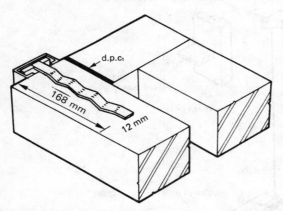

Figure 6.12 Pressed steel sub-frame fixed with corrugated adjustable lug

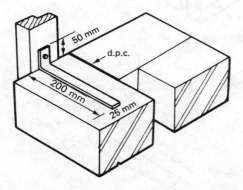

Figure 6.13 Timer door frame built in with galvanised metal strap

CAPILLARY ACTION

This is the term used to describe the peculiarity of water and other liquids of rising and travelling, vertically against the force of gravity.

All building materials that contain fine pores provide passageways for water to travel. Materials that contain minute pores allow the water to travel much further and also much more quickly. Figure 6.15 shows a simple experiment which may be carried out to illustrate the ability of water to rise against the force of gravity. The apparatus required is a water tray 50 mm in depth, and a simple form of stand, which is placed either across the tray or lengthways; the top rail of the stand has holes bored into it to accommodate up to six glass tubes, varying in diameter from 0.5 to 6 mm.

Experiment   Place coloured water in the tray to a depth of 32 mm, insert the glass tubes through their respective holes in the stand, allowing each tube to enter the water, and secure the tube, ensuring that its end is only just below the surface (figure 6.15).

The water will be seen to rise up the glass tubes, much higher and more quickly in the smaller diameter tubes. Figure 6.15 shows the glass tubes and the curve that may be obtained to denote the travel of the water, termed the 'curve of capillarity', proving that water has the ability to rise in narrow spaces. The fine pores in certain building materials will allow the water to travel unhindered, while materials such as no-fines concrete prevent capillary action from taking place.

To prevent capillarity taking place, many forms of construction are produced with 'anti-capillary grooves'. Figures 6.16—6.18 show these grooves formed between window sill and frame; water bars are also used to prevent this action and they should be inserted in all concrete and wooden sills.

WINDOW SILLS

Sills are formed at the bottom of window openings. The purpose of the sill is to prevent water entering the structure at this level; they also provide an attractive feature at the beginning of the opening. The materials used to form sills should be able to resist weather penetration such as bricks, tiles, stone, slate and concrete, although timber is also used.

The essential features required in sill construction are
(1) An adequate amount of slope or fall should be provided for the top surface of the sill; this is usually 6 mm per 100 mm of sill.

Figure 6.14 Capillary action

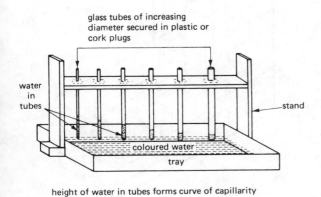

glass tubes of increasing
diameter secured in plastic or
cork plugs

water
in
tubes

stand

coloured water

tray

height of water in tubes forms curve of capillarity

Figure 6.15

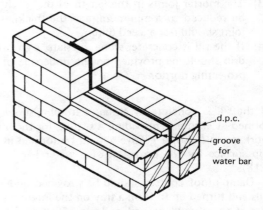

.d.p.c.

groove
for
water bar

Figure 6.16 Concrete or stone sill

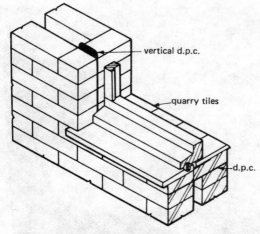

Figure 6.17 Slate sill

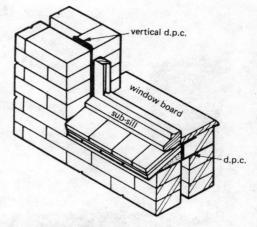

Figure 6.19 Quarry tile sill

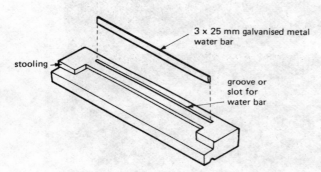

Figure 6.18 Concrete sill

Fixing Sills

When sills are formed with bricks it is essential to provide a horizontal gauge staff to determine mortar joint thickness and also to mark off the position of the bricks. When sills are fixed below eye level the line should be fixed to provide alignment of the top arris of the sill, but when the sill is above this height it is the bottom arris that is lined in (figures 6.21 and 6.22).

Brick and tile sills can be fixed before the frame is fitted, that is, during construction or after the openings have been formed and the frames fixed. The latter method often prevents damage occurring during construction.

(2) A generous amount of projection should be provided beyond the face of the wall. This should be at least 50 mm to allow water to fall clear of the wall face.

(3) The mortar joints in the length of the sill should be reduced to a minimum, and the thickness of joint should not exceed 6 mm.

(4) If the sill is concrete, stone, or slate, a groove or drip should be provided on the underside of the projecting portion.

If the sill is concrete or stone, a stooling should be formed at each end providing a seating for the brickwork reveal; also, a groove should be formed in the top surface for the provision of a water bar (figure 6.18).

Damp-proof courses should be provided under all sills and turned up to form a tray on the internal wall. The d.p.c. should extend at least 150 mm beyond each end of the sill (figures 6.16—6.21).

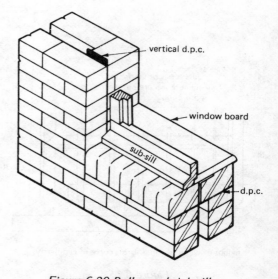

Figure 6.20 Bull-nose brick sill

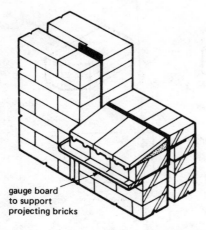

gauge board
to support
projecting bricks

Figure 6.21 Sill formed with roof tiles and projecting headers

Concrete and stone sills are usually fixed during construction. When several sills are required to be fixed on one elevation, the two end sills should be fixed first and the intermediate sills then lined in. A solid bed of cement mortar is formed at each end of the sill only, that is, for a length of approximately 150–200 mm, and the remaining portion of the sill is left with an open joint until the structure is completed. This open joint is later filled with a semi-stiff mix of mortar: this practice is carried out to counteract any discrepancy in settlement, and to prevent fracturing from occurring if a solid bed was used for the whole length of the sill.

Water Bars

These are formed of galvanised metal bar, 25 x 3 mm, fixed into the groove formed in the top surface of the structural sill and the groove in the underside of the timber sub-sill. The bar is sealed into position with bituminous mastic and prevents capillarity occurring between the two sills (see figure 6.18).

Protection of Sills   When sills are fixed during construction it is necessary to provide protection against falling objects, abrasion, weather and staining. Boards placed on felts or quilts are often used, incorporating lightweight plastic sheeting, but whatever method is used it is essential that it should be able to resist displacement and should be capable of protecting the sill from staining.

REVEALS

There are three methods of forming reveals to openings each of which can be carried out with normal bonding arrangements. The types of reveal are

(1) plain reveal
(2) recessed reveal
(3) splayed reveal.

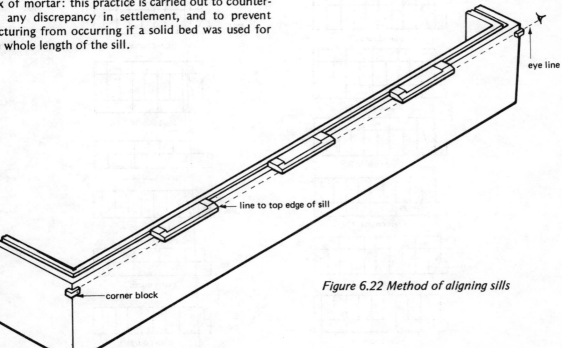

eye line

line to top edge of sill

corner block

Figure 6.22 Method of aligning sills

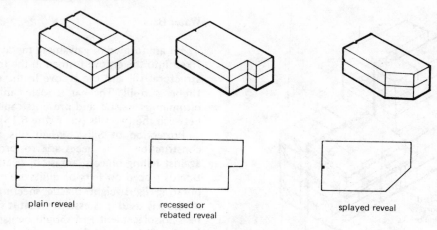

plain reveal

recessed or
rebated reveal

splayed reveal

Figure 6.23

Plain Reveal

This is used for normal openings in walls; the reveal is at right-angles to the wall face. This type of reveal can be formed in both cavity and solid walls (figure 6.23).

Recessed Reveal

This reveal is formed with a recess or rebate, which should be half or quarter brick; cavity or solid walls may contain this type of reveal. The advantage

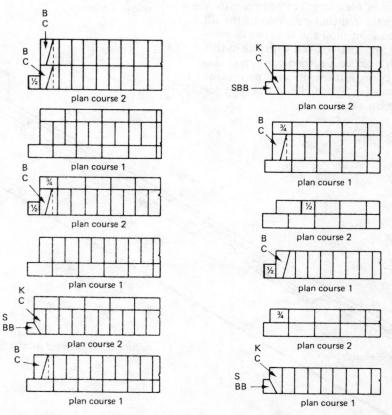

Figure 6.24 Rebated or recessed reveals in English bond

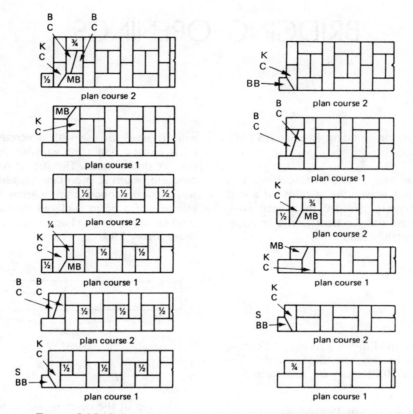

Figure 6.25 Rebated or recessed reveals in Flemish bond

offered is improved security for fixing of frames, and there is less likelihood of weather penetration, since the recess forms an abutment for the frame (figures 6.23—6.25).

Splayed Reveal

This type of reveal is not often used, its purpose being to infuse more light into the interior of the building. It can be used for solid and cavity walls, considerable skill being required in the bonding arrangement (figure 6.23).

It is possible to obtain a combination of splayed and recessed, but difficulty in bonding often occurs and unless there is an advantage to be gained this type of reveal is not often used.

Bonding of a rebated or recessed reveal in a cavity wall is illustrated in figure 6.26.

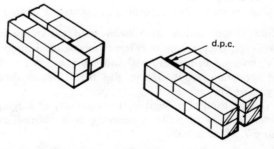

Figure 6.26 Bonding rebated or recessed reveal in cavity wall

7

BRIDGING OPENINGS

Part A1/2 (C29) of the Building Regulations 1985 states:

> The number, size and position of openings and recesses should not impair the stability of a wall or the lateral support afforded by a buttressing wall to a supported wall. Construction over openings must be adequately supported.

LINTELS

Timber Lintels

These were regularly used in the past and, depending on the span and the loading, are still very useful. The advantages of timber lintels can be summarised as follows

(1) they are light and thus easy to place
(2) they are cheap in comparison with other materials
(3) fixings are simple for architraves, pelmets, curtain rails, etc.

Disadvantages include

(1) the possibility of sagging under load when inadequately sized
(2) spans are limited according to the sizes available.

While timber lintels have been used externally in dwellings, they are more often than not used internally, covered in plasterboard and skimmed, not only for appearance, but also to give them the required degree of fire resistance.

One method of preventing the possibility of sagging under load is to build a relieving arch above (see figure 7.54, page 60).

Reinforced Concrete Lintels

Iron bars were first used to reinforce concrete in about 1870. After the beginning of the twentieth century, mild-steel bars were used for this purpose,

and are ideal for use with concrete since both materials expand and contract with temperature variation at similar rates. The use of reinforcing bars in concrete beams is important because of the inherent weakness of concrete in tension (stretching). For example, if a beam is placed over an opening it will tend to sag under load and stresses will occur in the beam (figure 7.1).

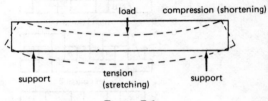

Figure 7.1

The top of the beam is said to be in compression (shortening) and the underside is in tension (stretching), and since concrete is known to be strong in compression and weak in tension, reinforcing bars are built into the tensile zone.

Probably the simplest way to understand tension and compression is to take a 1 m length of 75 x 15 timber and introduce saw cuts along both edges (see figure 7.2). Support this on edge at each end and

sawcuts top and bottom

Figure 7.2

apply a load at the centre. The saw cuts on the top will close up and those on the underside will open, illustrating compression and tension (figure 7.3).

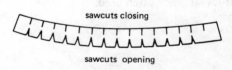

sawcuts closing

sawcuts opening

Figure 7.3

The centre of the beam does not lengthen or shorten and is therefore in neither compression nor tension, as the stress diagram in figure 7.4 shows. This point is known as the neutral axis.

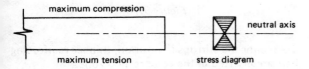

Figure 7.4

As already stated, concrete is stronger in compression than in tension, ten times as strong in fact, and thus it is the edge that is in tension that has to be reinforced. In simple beams the bars are usually inserted 25 mm from the bottom. This concrete cover is necessary to prevent any possibility of the bars rusting and to bond thoroughly with the concrete. Also, to prevent any possibility of slip inside the beam the bars are usually hooked at their ends, as shown in figure 7.5.

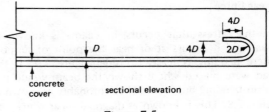

Figure 7.5

Figure 7.6 shows some of the situations in which concrete beams may be placed, and opposite these are the shapes the beams will be likely to take when loaded (this is exaggerated). In each case the edge that is tending to stretch is the tensile edge and the reinforcing steel should be here. The top of a lintel should always be clearly marked while the concrete is still green (not fully hardened), unless this will be perfectly obvious to the user as in the case of boot lintels (figure 7.11). If this is not done the lintel may be placed in position upside down, in which case the bars would be in the compression zone and the lintel would be liable to fail in tension.

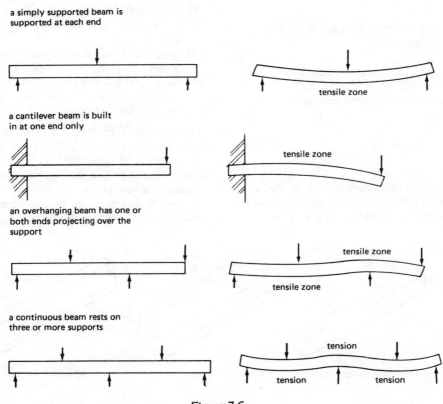

Figure 7.6

Shear Force

Another stress that occurs in beams is known as shear and it occurs at or near the point of support where bending is least. Consider figure 7.7. If the load were placed where shown the beam would not fail in bending but would shear probably as shown in figure 7.8. This is known as diagonal shear which can be counteracted in two ways

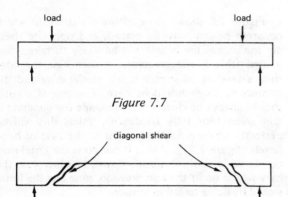

Figure 7.7

Figure 7.8

(1) by cranking up alternate bars across the expected failure zone (figure 7.9)

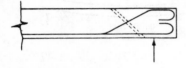

Figure 7.9

(2) by including stirrups (steel wires) in the shear zone, decreasing towards the centre of the beam where shear force is at its least (figure 7.10).

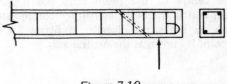

Figure 7.10

To anchor the stirrups in position, thinner reinforcing bars are inserted in the compression zone.

Concrete lintels are very versatile and can be cast to the exact size and shape required. Some of the possibilities are shown in figure 7.11.

Another method of classification of concrete members is to describe them as precast or *in situ*.

Precast Lintels

These are cast at a convenient level and can be raised and placed into position when required, followed by immediate loading, thus causing no hold-up in the work of the bricklayers. They can be obtained from a precast manufacturer's yard or cast on site if time permits. Figure 7.12 shows a convenient mould for casting several small lintels. The formwork used must be

(1) sufficiently strong to withstand vibration
(2) sufficiently rigid: no deflection must occur while casting
(3) grout-tight: no cement must seep through the joints
(4) easily erected and dismantled; screws are preferable to nails and, if the latter are used, double-headed nails are best.

The bars must be inserted in the formwork in the required position and perhaps the best method of doing this is to stand the bars on concrete or plastic spacer blocks to give the required cover (figure 7.13).

When spacer blocks are not available, the bottom 25 mm of concrete can be poured and consolidated and the bars inserted at this point, after which the

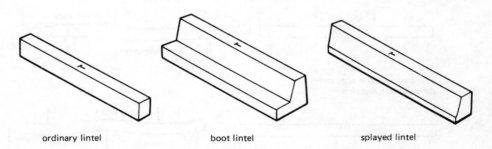

ordinary lintel   boot lintel   splayed lintel

Figure 7.11

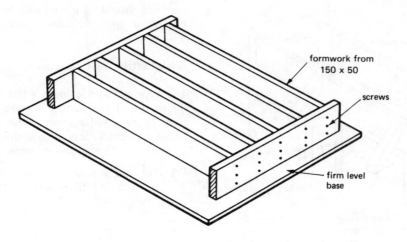

suitable mould for casting four
900 × 100 × 150 lintels

Figure 7.12

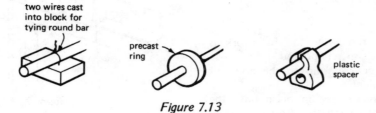

Figure 7.13

rest of the concreting is completed. This method is often used for small lintels, but for work of more importance spacer blocks are to be preferred.

Wood pads for fixings can be cast in a lintel by tacking the wedge-shaped pads inside the formwork in the correct position and pouring the concrete around these. When the formwork is stripped the fixings will be firmly cast into the lintel.

Perhaps the only problem with precast concrete lintels is due to their mass (weight). If lifting equipment is not available or headroom is limited, difficulties in raising and placing could arise.

In Situ Lintels

The bricklayer completes work to the top of the lintel, leaving the required bearing at each end (figure 7.14).

Formwork can now be fixed firmly in position, oiled for ease of striking and the concrete poured; the bars are inserted in the required position as previously explained. The formwork is struck (removed) when the concrete has hardened sufficiently. See figure 7.15.

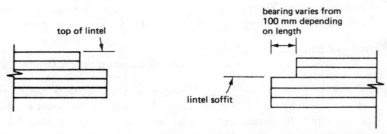

Figure 7.14

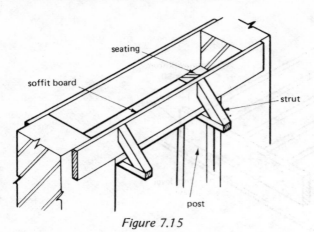

Figure 7.15

The advantages of *in situ* concrete lintels are

(1) they can be cast to any required shape and size
(2) no lifting equipment is necessary
(3) they are useful for providing support for large-span brick lintels (see figure 7.17).

Disadvantages are

(1) bricklaying must stop while the formwork is being fixed and the concrete poured
(2) moving the wet concrete up to the point of placing is inconvenient
(3) the formwork has to stay in place until the concrete has gained strength, although certain members can be removed to allow bricklaying to continue.

Sizing Concrete Lintels

It is preferable to have a beam designed to suit a specific purpose, but for simply supported beams without complex loading the following rule-of-thumb method is useful.

Length	Lintel depth	Bar dia.
up to 1200	2 courses	12 mm
up to 1600	3 courses	16 mm
up to 2000	3 courses	20 mm
up to 2500	4 courses	20 mm
up to 3000	4 courses	25 mm

One reinforcing bar is used per half brick thickness of walling.

Concrete Mixes

1:2:4 is the leanest mix for concrete lintels, but richer mixes may be used if considered necessary. The concrete should be placed into formwork as dry as is consistent with workability and cured for at least 3 days.

Brick Lintels

This type of lintel is simply a flat arch and because of the nature of its structure is a weak form of construction. Therefore unless some other form of support is provided the span should be limited to 600 mm.

Methods of increasing the span are as follows.

(1) Build on a mild-steel angle iron which rests on supports at either side of the opening (figure 7.16).

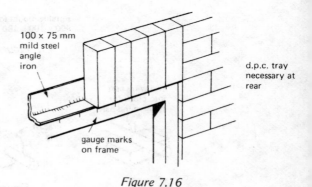

Figure 7.16

(2) Build the lintel on the frame or a temporary support and leave wall ties projecting at the back. An *in situ* concrete lintel is then cast behind this (figure 7.17).
(3) Build on a reinforced concrete boot lintel (figure 7.18). A d.p.c. is necessary.
(4) Provide extra support by means of a galvanised steel lintel (Catnic, Dorman Long, Birtley, etc.) Figure 7.19 illustrates a Birtley combined galvanised steel angle. No d.p.c. is required, but it is generally considered good practice to include one.
(5) Build with perforated bricks threaded on to a steel reinforcing bar (rarely used) (figure 7.20). Some support must be provided during building and spans should be limited.

British Standard Beams (figure 7.21)

Where openings are wide, loads to be carried are great and the depth of the beam is limited BS beams may be preferred. For example, if a loadbearing wall is removed from the ground storey in a house, a BS

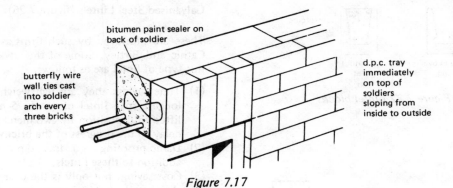

bitumen paint sealer on
back of soldier

butterfly wire
wall ties cast
into soldier
arch every
three bricks

d.p.c. tray
immediately
on top of
soldiers
sloping from
inside to outside

Figure 7.17

beam is often the choice for supporting the load since
it is lighter and easier to handle and install than a
concrete beam. Many sizes are available and the loads
they will support are tabulated in the British Constructional Steelwork Association Handbook. When
installed in dwellings they must be boxed in to give
the required fire resistance.

Use of Lintel Blocks

Where blocks are used for fair-faced walling it is
sometimes possible to avoid the use of a concrete
lintel for bridging openings by using specials, inside
which a lintel can be cast. For example, Forticrete

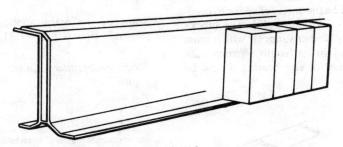

Figure 7.18

Figure 7.19

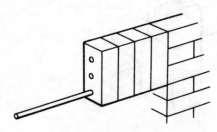

Figure 7.20

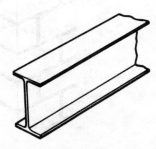

Figure 7.21

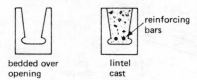

Figure 7.22 Lintel blocks

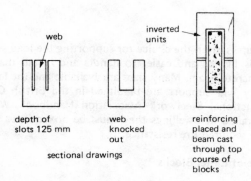

Figure 7.23 Bond beam units

Ltd make various specials, including lintel blocks for small openings and bond beam units for spans exceeding 1.800 m (see figures 7.22 and 7.23). The blocks are laid over a temporary timber support, which must be adequately supported in its length (figures 7.24 and 7.25) and the amount of end bearing should be not less than 190 mm.

Galvanised Steel Lintels (figure 7.26)

These are produced by such firms as Dorman Long, Catnic and Birtley. Some of the advantages of using this type of lintel are as follows.

(1) Time saving: they are lightweight and even the longest span lintel of nearly 5 m can easily be lifted into position by two men. Thus there is no delay to the progress of the brickwork.
(2) Damp-proofing: no tray d.p.c. is required in addition to these lintels.
(3) Cost saving: not only is the cost of a tray d.p.c. eliminated, but in some cases one course of blocks or up to three courses of bricks are saved up to the full length of the lintel.
(4) High quality: the strength of the steel combined with a hot-dip galvanising for protection against rust and corrosion ensures a permanent, sound job.
(5) Versatility: the wide range of sizes and types available means that galvanised steel lintels are available to suit most domestic requirements. They are suitable for all types of construction such as blockwork, brickwork, reconstructed stone, etc. Expanded steel mesh is welded to any faces intended for plastering.
(6) Free advice: this is available from the technical department of each of these firms.

Applications

Some of the many applications of these lintels are shown in figures 7.27–7.31 in section only.

Single or Standard Lintels   These are illustrated in figure 7.27.

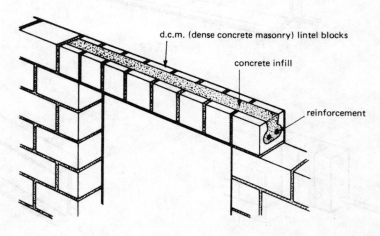

Figure 7.24

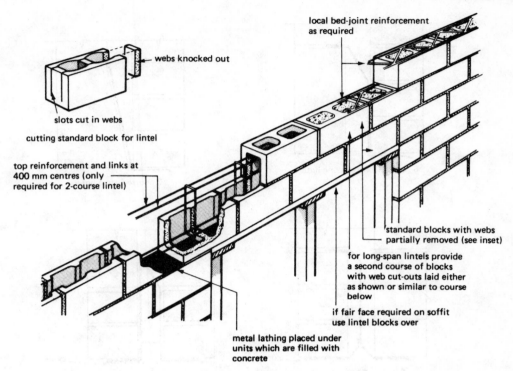

webs knocked out

slots cut in webs

cutting standard block for lintel

top reinforcement and links at
400 mm centres (only
required for 2-course lintel)

local bed-joint reinforcement
as required

standard blocks with webs
partially removed (see inset)

for long-span lintels provide
a second course of blocks
with web cut-outs laid either
as shown or similar to course
below

if fair face required on soffit
use lintel blocks over

metal lathing placed under
units which are filled with
concrete

Figure 7.25 Bond beam and large-span lintel

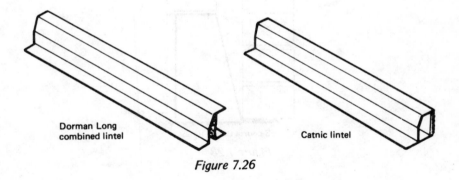

Dorman Long
combined lintel

Catnic lintel

Figure 7.26

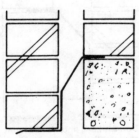

2-course Birtley lintel

Figure 7.27

Combined Lintel These carry both leaves of a cavity wall; they are illustrated in figure 7.26.

Some Other Applications These are illustrated in figures 7.29—7.31.

These are just a few of the possible applications of galvanised steel lintels. They can also be used for

(1) openings in 100 mm and 150 mm-wide single walls
(2) fixing to concrete floor beams to carry the outer skin of cavity walls

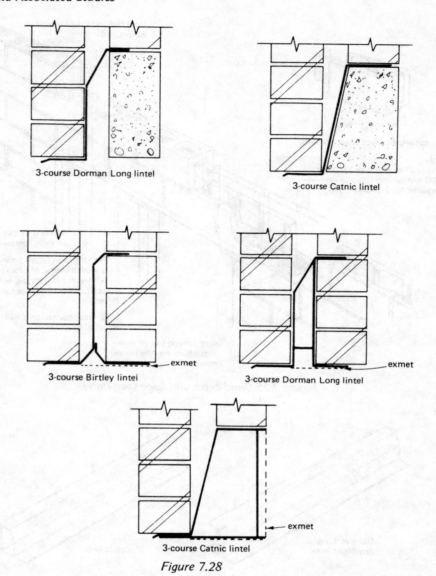

3-course Dorman Long lintel

3-course Catnic lintel

3-course Birtley lintel — exmet

3-course Dorman Long lintel — exmet

3-course Catnic lintel — exmet

Figure 7.28

Catnic lintel for cavity walls
with 150 mm inner skin

Figure 7.29

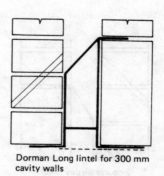

Dorman Long lintel for 300 mm
cavity walls

Figure 7.30

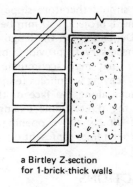

a Birtley Z-section
for 1-brick-thick walls

Figure 7.31

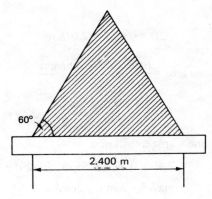

Figure 7.32

(3) at eaves level to support roof construction
(4) over door openings in internal walls 75 mm and 100 mm wide (50 mm only in depth of lintels).

Advice

Catnic Components Ltd offer the following advice to the users of galvanised steel lintels

(1) do not use damaged lintels
(2) always provide 150 mm end bearings
(3) do not apply point loads without consultation
(4) do not apply concrete floor loads without first checking suitability
(5) live loads not to exceed 50 per cent of total loading
(6) in cavity walling the brickwork and blockwork must be raised uniformly during and after building-in these lintels
(7) for external walls, not more than half the load should be carried by the outer leaf and the wall must be restrained laterally near the top.

Loads on Lintels

The load carried by a lintel in a properly bonded brick wall is incorporated in an equilateral triangle, as shown in figure 7.32. It is thus a simple task for the structural engineer to design a suitable beam, whether of timber, concrete or steel, once he has calculated the total loading.

Example 7.1

To find the mass (weight) supported by the lintel shown, assuming that the brickwork above is 220 mm thick and the density of the brickwork is 1800 kg/m^3.

$$\text{total mass} = \text{volume} \times \text{density}$$

and

$$\text{volume} = \text{shaded area} \times \text{thickness}$$

$$\text{shaded area} = \frac{\text{base} \times \text{vertical height}}{2}$$

In figure 7.33, to find the vertical height: by Pythagoras

$$AC^2 = CD^2 + AD^2$$

therefore

$$CD^2 = AC^2 - AD^2$$
$$= 2.4^2 - 1.2^2$$
$$= 5.76 - 1.44$$
$$= 4.32$$

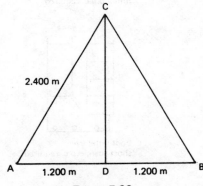

Figure 7.33

therefore

$$CD = \sqrt{4.32} = 2.078 \text{ m}$$

$$\text{shaded area} = \frac{\text{base x vertical height}}{2}$$

$$= \frac{2.4 \times 2.078}{2} = \frac{4.99}{2} = 2.49 \text{ m}^2$$

volume = area x thickness

$$= 2.49 \times 0.22 = 0.55 \text{ m}^3 \text{ (to 2 d.p.)}$$

mass = volume x density

$$= 0.55 \times 1800$$

$$= 990 \text{ kg (nearly 1 tonne)}$$

DAMP-PROOF COURSES

Whenever a cavity is bridged otherwise than by
(a) a wall tie; or
(b) a bridging which occurs at the top of a wall in such a position that it is protected by a roof, then

a d.p.c. or flashing shall be inserted in such a manner as will prevent the passage of moisture from the outer leaf to the inner leaf of the wall.

This d.p.c. should therefore slope from inside to outside and must project 150 mm beyond each end of the lintel. Depending on the length of the lintel it may also be necessary to provide a means of escape for any water that may accumulate on the d.p.c. This is done by leaving weepholes (empty cross joints) every four bricks along the face of the wall on the course above the d.p.c. Some examples of the requirements of d.p.c.s above openings are shown in figure 7.34. Broken lines represent d.p.c.s.

BEAM REACTIONS

The design of a beam, whether of timber, steel or reinforced concrete, is a simple task for a structural engineer once the loads to be superimposed are known. There is, however, a further consideration and that is the reactions at the supports; for example, would a 75 mm block wall support a concrete lintel over a 4 m span?

There are two types of load on beams: point loads and uniformly distributed loads.

Point Loads

Loads are described as point loads where a load occurs on a beam in one place only (there may be more than one point load). An example of a point load is shown in figure 7.35. Point loads behave as shown in figure 7.36.

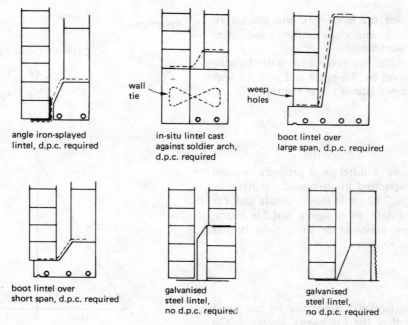

angle iron-splayed
lintel, d.p.c. required

wall
tie

in-situ lintel cast
against soldier arch,
d.p.c. required

weep
holes

boot lintel over
large span, d.p.c. required

boot lintel over
short span, d.p.c. required

galvanised
steel lintel,
no d.p.c. required

galvanised
steel lintel,
no d.p.c. required

Figure 7.34

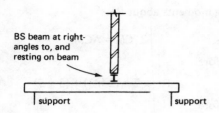

BS beam at right-angles to, and resting on beam

| support | support |

Figure 7.35

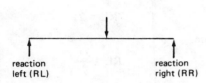

reaction left (RL)   reaction right (RR)

Figure 7.36

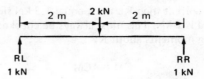

RL 1 kN   RR 1 kN

Figure 7.39

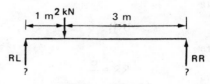

Figure 7.40

Uniformly Distributed Loads (UDLs)

Loads are described as uniformly distributed where the load is dispersed over part or the whole of the beam. An example of a UDL is shown in figure 7.37. UDLs are usually shown as in figure 7.38.

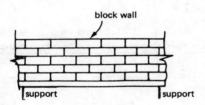

block wall

| support | support |

Figure 7.37

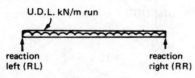

U.D.L. kN/m run

reaction left (RL)   reaction right (RR)

Figure 7.38

Calculation of Beam Reactions

Point Loads

If a load of 2 kN were placed at the centre of a beam the supports would carry 1 kN each (figure 7.39), but if the load were placed nearer to one end the reaction at each support would not be so obvious (figure 7.40).

To find the reactions at the supports it is necessary to take moments about each end in turn.

The Principle of Moments If a beam is in equilibrium under the action of a number of forces, then the clockwise moments about each end must equal the anticlockwise moments.

Example 7.2

In figure 7.41, taking moments about RL

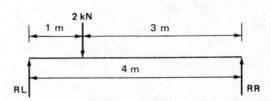

Figure 7.41

Clockwise moments (CM) = anticlockwise moments (ACM)

Therefore

$$2 \times 1 = 4 \times RR$$

Therefore

$$2 = 4RR$$

Therefore

$$RR = \frac{2}{4}$$

$$= 0.5 \text{ kN}$$

If this is correct then RL must equal 1.5 kN since the total load is 2 kN. This must always be checked.

Taking moments about RR

$$CM = ACM$$

Therefore

$$RR \times 4 = 2 \times 3$$

Therefore

$$4RR = 6$$

Therefore

$$RR = \frac{6}{4}$$

$$= 1.5 \text{ kN}$$

Note When taking moments about RL you are finding RR and vice versa.

Example 7.3

Where a beam carries two point loads the procedure is similar, the magnitude of the point loads being multiplied by their distances from the support.

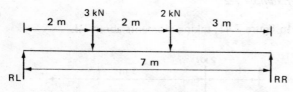

Figure 7.42

In figure 7.42, taking moments about RL

$$CM = ACM$$

Therefore

$$(3 \times 2) + (2 \times 4) = 7RR$$

Therefore

$$6 + 8 = 7RR$$

Therefore

$$14 = 7RR$$

Therefore

$$RR = \frac{14}{7}$$

$$= 2 \text{ kN}$$

Taking moments about RR

$$CM = ACM$$

Therefore

$$RL \times 7 = (2 \times 3) + (3 \times 5)$$

Therefore

$$7RL = 6 + 15$$

Therefore

$$7RL = 21$$

Therefore

$$RL = \frac{21}{7}$$

$$= 3 \text{ kN}$$

These reactions add up to 5 kN, which is the total load.

UD Loads

The total UDL is calculated and this can then be considered as a point load acting through its centre of gravity.

Example 7.4

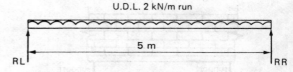

Figure 7.43

In figure 7.43

$$\text{total UDL} = \text{length} \times \text{load/m run}$$

$$= 5 \times 2$$

$$= 10 \text{ kN}$$

The diagram is thus as shown in figure 7.44, the reactions both being 5 kN.

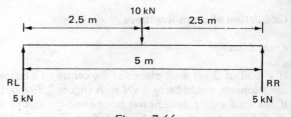

Figure 7.44

Example 7.5

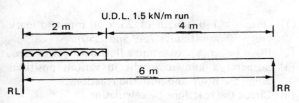

Figure 7.45

In figure 7.45

$$\text{total UDL} = \text{length} \times \text{load/m run}$$
$$= 2 \times 1.5 \text{ kN}$$
$$= 3 \text{ kN}$$

As explained, this is now treated as a point load acting at its centre of gravity and the diagram becomes

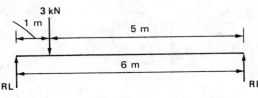

Figure 7.46

as shown in figure 7.46. Taking moments about RL

$$CM = ACM$$

Therefore

$$3 \times 1 = 6 \times RR$$

Therefore

$$3 = 6RR$$

Therefore

$$RR = \frac{3}{6}$$
$$= 0.5 \text{ kN}$$

Taking moments about RR

$$CM = ACM$$

Therefore

$$6RL = 3 \times 5$$

Therefore

$$6RL = 15$$

Therefore

$$RL = \frac{15}{6}$$
$$= 2.5 \text{ kN}$$

Check

$$0.5 + 2.5 = 3 \text{ kN (the point load)}$$

The final example shows a beam carrying a point load and a UDL (figure 7.47). The UDL is 1.5 kN/m

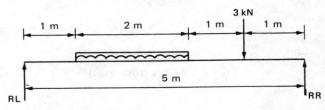

Figure 7.47

run, and is treated as a point load of 3 kN (2 x 1.5) acting at its centre of gravity, that is, 2 m from RL. Thus the load on the beam becomes as shown in figure 7.48.

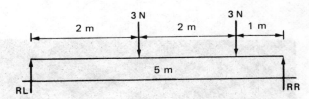

Figure 7.48

Example 7.6

Taking moments about RL

$$CM = ACM$$
$$(3 \times 2) + (3 \times 4) = 5RR$$

Therefore

$$6 + 12 = 5RR$$

Therefore

$$18 = 5RR$$

Therefore

$$RR = \frac{18}{5}$$
$$= 3.6 \text{ kN}$$

Taking moments about RR

$$CM = ACM$$

$$5RL = (3 \times 1) + (3 \times 3)$$

Therefore

$$5RL = 3 + 9$$

Therefore

$$5RL = 12$$

Therefore

$$RL = \frac{12}{5}$$

$$= 2.4 \text{ kN}$$

Check

$$3.6 + 2.4 = 6 \text{ kN (total load)}$$

Experiment to Observe the Reactions of a Simply Supported Beam

(1) Obtain a 1 m length of 25 x 50 mm and mark every 50 mm (figures 7.49 and 7.50).
(2) Place beam as shown and adjust balances to zero.
(3) Suspend a known weight in various positions along the beam and note the reactions.
(4) Check the reactions by calculation.
(5) Repeat with other weights.

Note   The sum of the reactions is always equal to the total load.

ARCHES

A more decorative but much more expensive method of bridging an opening is to use one of the many

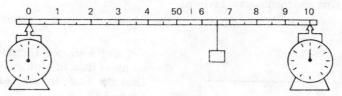

Figure 7.49

Figure 7.50

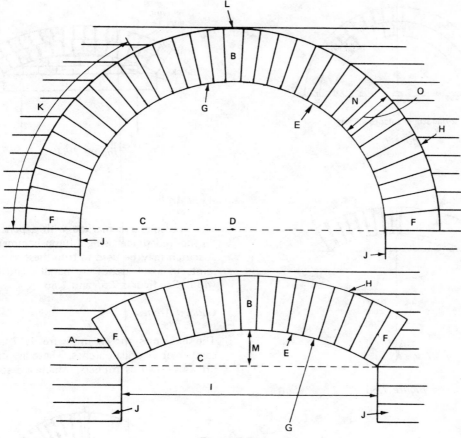

Figure 7.51

forms of arch available. The soldier or flat arch has already been dealt with since this is really a brick lintel.

Terms Used

The terms used are as follows (see figure 7.51).

A, *skewbacks*, the inclined or splayed surfaces from which certain types of arch spring.

B, *key brick*, the central brick in the arch face, always placed last.

C, *springing line*, a horizontal line joining the two springing points and equal in length to the span.

D, *striking point*, the centre of the circle of which the intrados and extrados are arcs.

E, *soffit*, the underside of the arch.

F, *springers*, the lowest voussoirs on either side of an arch.

G, *intrados*, the inner or bottom curve of the arch.

H, *extrados*, the external or top curve of the arch.

I, *span*, the distance between the abutments.

J, *abutments*, the brickwork on either side supporting the arch.

K, *haunch*, the lower third of the arch.

L, *crown*, the highest point on the extrados.

M, *rise*, the vertical distance between the springing line and the highest point on the intrados.

N, *voussoirs*, the bricks used for building the arch.

O, *face depth*, the dimension between intrados and extrados.

Classification

Arches may be classified in any of the following ways

(1) rough, axed or gauged
(2) by their shape, that is, segmental, semicircular, Gothic, etc. (figures 7.52 and 7.53)
(3) by the number of centres from which the arch is struck: 1, 2, 3, 4 or 5 centred.

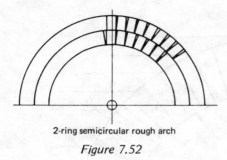

2-ring semicircular rough arch

Figure 7.52

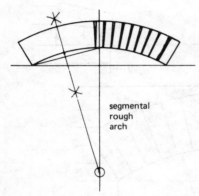

segmental
rough
arch

Figure 7.53

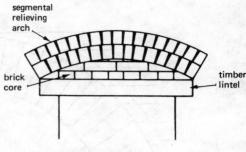

segmental
relieving
arch

brick
core

timber
lintel

Figure 7.54

Rough Arches

These are turned using uncut voussoirs, the curvature being obtained by making the joints V-shaped. No setting out or traversing (see pp. 62 and 63) is necessary for rough arches. All that is required is to set the support correctly, and to step off the voussoirs on the support, keeping the joints as tight as possible on the intrados; the width of joints on the extrados will depend on the curvature of the arch. It is important that there are an odd number of voussoirs if the arch has a curved soffit, otherwise there will be no key brick.

Rough arches are generally used where the main objective is to span the opening and the appearance is of secondary importance. They are often used in two half-brick rings as relieving arches to take the load off a timber lintel (figure 7.54).

In a relieving arch the brick core is cut to follow the soffit curve using a plywood templet to show the desired shape. The skewbacks are cut and bedded and the arch is turned over a sand bed, which allows for slight movement to occur without putting pressure on the core. The joint between the core and the arch soffit is then pointed up with mortar.

Axed Arches

The voussoirs are marked with a pencil and cut to the desired shape using a lump hammer and bolster. A scutch may be used to trim these as necessary. The joints in axed arches are parallel and 8–10 mm in width. See figures 7.55 and 7.56.

Gauged Arches

The examples shown in figure 7.57 are known as camber or Georgian arches. These are constructed of bricks known as rubbers, which are soft enough to

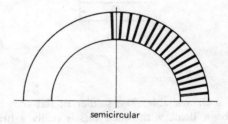

semicircular

Figure 7.55

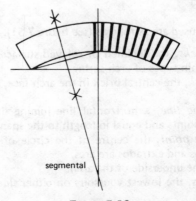

segmental

Figure 7.56

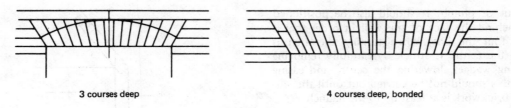

| 3 courses deep | 4 courses deep, bonded |

Figure 7.57

be cut with a bow saw and rubbed on a stone to the exact shape required.

The voussoirs are then set in lime putty with 3 mm joints and, as shown in figure 7.57, the rubbers are obtainable oversized (250 x 125 x 75).

Temporary Support of Arches

An arch is turned over a centre or turning piece which is made by the joiner. The type of arch being built often determines the type of support.

Turning Pieces (figure 7.58)

This is suitable for an arch of low rise such as a camber or segmental arch. It is simpler and quicker to construct than a centre, being cut from a solid piece of timber with the top edge cut to the desired curvature.

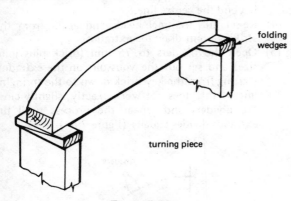

Figure 7.58

Centres (figure 7.59)

Figures 7.59 and 7.60 show centres suitable for semicircular and segmental arches respectively. Closed laggings can be used to provide a more accurate support for gauged brick arches, open laggings being suitable for most axed or rough arches.

When the bricklayer has completed the work up to the springing line the turning piece or centre has to be fixed safely and firmly in position and in such

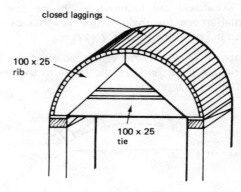

Figure 7.59

a way that it can easily be struck (removed) on completion.

Support can be provided by adjustable steel or timber props, depending on the availability, and these are placed in position about 25 mm below the base of the centre. The centre is then positioned by means of a pair of folding wedges on each support (figure 7.58).

The centre must be level across the bottom and also across the top centre lagging or the arch may be

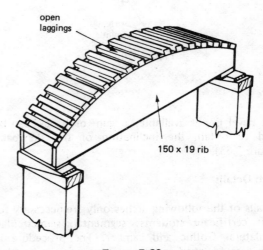

Figure 7.60

turned out of plumb. It should also be positioned slightly behind the face of the wall in order that it will not foul the line. When turning the arch is complete the centre is struck by carefully removing the folding wedges, lowering the centre and easing out, but this should not be carried out until the surrounding brickwork is at least up to the haunch.

Skewbacks

Where skewbacks are required, as in the case of segmental arches, the angle of the skewback can be obtained in one of the following ways

(1) place a brick on the centre adjacent to the springing point and set the bevel to the angle produced (figure 7.61)

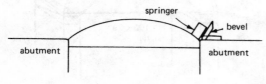

Figure 7.61

(2) pull a line from the striking point through the springing line at the point of support and set the bevel to this angle (figure 7.62).

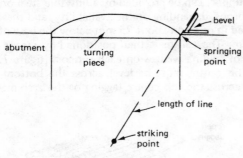

Figure 7.62

Note

If a bevel is not available, a 'gun' or 'stock' can be used to obtain the inclination of the skewback (figure 7.63).

Arch Details

Details of the following arches only are necessary for craft certificate students: segmental, semicircular, equilateral Gothic and camber. The procedure in each case for axed and gauged arches is similar, that

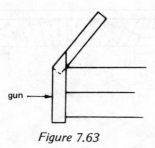

Figure 7.63

is, setting out, traversing, cutting and building. Requirements for setting out and traversing are as follows: a sheet of plywood sufficiently large to accommodate just over half the arch, trammel heads, dividers, tape, saw, plane, two traversing rules (pieces of wood 350 x 50 x 10 mm approximately), straight-edge and a piece of ply for a templet.

Axed Semicircular Arch (figure 7.55)

Setting Out

(1) Rest the sheet of ply on a flat, level surface at a convenient height.
(2) Set out the springing line and at right-angles to this the centre line.
(3) Open the trammel heads to half the span and from the striking point draw the intrados to just beyond the centre line.
(4) Re-set the trammel heads and again from the striking point draw the extrados.
(5) Open the dividers to 75 mm (brick plus joint size) and set out the voussoirs on the extrados, starting from the key brick down to the springing line. If this does not work exactly, slightly close the dividers and repeat the process until the extrados divides equally (figure 7.64).

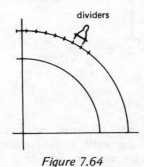

Figure 7.64

(6) Draw in the voussoir lines from the striking point extending below and above the arch face (figure 7.65).

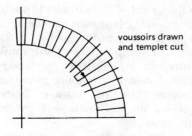

Figure 7.65

(7) Place the piece of ply over a voussoir and mark
   the voussoir lines on here.
(8) Cut the ply to the marked shape to form the
   templet (figure 7.65).
(9) Replace the templet over the voussoir and mark
   on the position of the intrados.

Traversing

(1) Mark the traversing rules 1 and 2 and place
   rule 1 against the left-hand side of the key
   brick (figure 7.66).

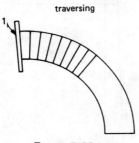

Figure 7.66

(2) Place the templet against this rule with the
   intrados position on the templet coinciding
   with that on the drawing (figure 7.67).
(3) Place rule 2 against the templet (figure 7.68).
(4) Remove rule 1 and the templet and place rule 1
   against the left side of rule 2 (figures 7.69 and
   7.70).

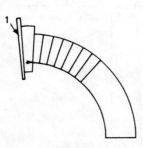

Figure 7.67

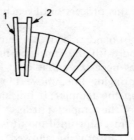

Figure 7.68

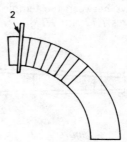

Figure 7.69

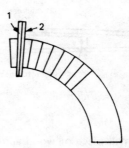

Figure 7.70

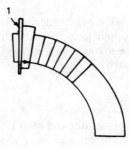

Figure 7.71

(5) Remove rule 2 and place the templet against
   the right edge of rule 1 (figure 7.71).

(6) Continue this process until reaching the springing line.

(7) If the right-hand side of the templet does not correspond with the springing line, adjust the templet as necessary, that is, if finishing short of the springing line, move the intrados line further up the templet; if finishing past, plane the edge of the templet a little.

(8) Continue traversing until the fit is correct.

(9) The joint can now be allowed for by placing the templet between the two traversing rules and sliding the templet upwards, tight against one of the rules, until the required joint thickness shows between the templet and the other rule (figures 7.72 and 7.73).

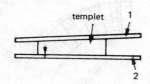

Figure 7.72

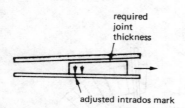

Figure 7.73

(10) Tack a small piece of wood across the templet at the new intrados position.

Cutting Voussoirs

The templet is set centrally against the face of the brick, which is then marked with a pencil. All the bricks should be marked in this way, then cut and trimmed as required.

Building

(1) The opening is set out and built plus a 3 to 5 mm tolerance. A pinch rod is cut to this overall dimension and abutments are checked with the pinch rod as work proceeds.

(2) The abutments are erected to the springing line and the centre is erected clear of the face line and adjusted with folding wedges.

(3) The brickwork is erected either side of the arch

to the height of the extrados if possible. If this is not possible, dead men may be built either side to line in the face of the arch.

(4) Step out the positions of the voussoirs on the centre, fix a line across the face position and fasten a short length of line to the striking point.

(5) Start bedding the voussoirs from each side, checking the bed joints with the line from the striking point.

(6) Place the key brick last, ensuring full joints either side.

(7) Do not strike the centre until the surrounding brickwork covers the haunches. Then, finally, point up the soffit.

Axed Segmental Arch (figure 7.56)

Setting Out

(1) Measure the rise and the span and set out just over half the span AB on a sheet of ply (figure 7.74).

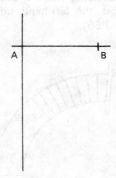

Figure 7.74

(2) Draw the centre line of the arch perpendicularly to the springing line and mark off the rise AC (figure 7.75).

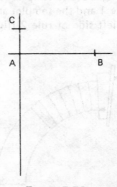

Figure 7.75

(3) Join BC and vertically bisect, carrying this down to intersect with CA produced to D (figure 7.76).

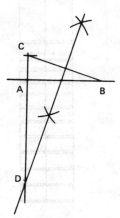

Figure 7.76

(4) D is the striking point and the trammel heads are opened a distance equal to CD. The intrados can now be drawn from B to a point just past C (figure 7.77).

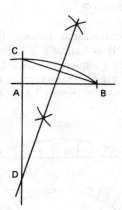

Figure 7.77

(5) Open the trammel heads further to give the face depth of the arch and, with D again as the striking point, draw the extrados (figure 7.78).

(6) The angle of skewback is obtained by drawing a line from D through B to intersect with the extrados (figure 7.79).

(7) Open the dividers 75 mm (brick plus joint) and mark the position of the key brick on the extrados. From this point, setting out, traversing and cutting are as for the axed semicircular arch.

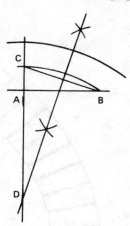

Figure 7.78

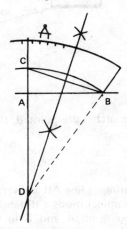

Figure 7.79

Building

(1) Cut and bed the skewbacks and erect the brickwork either side of the arch to the height of the top of the extrados.

(2) Set the centre and mark on the voussoir positions as far as possible.

(3) Fix a cross member in the opening so that a nail can be put in at the striking point, to which a short length of line can be fixed to check the joints between the voussoirs as the arch is turned.

(4) Turn the arch from both sides, placing the key brick last.

(5) Strike the centre and point up the soffit.

Axed Equilateral Gothic Arch (figure 7.80)

At this point, an important rule appertaining to arches should be made clear. This is

where there is a point there is a joint,
where there is a curve there is a key

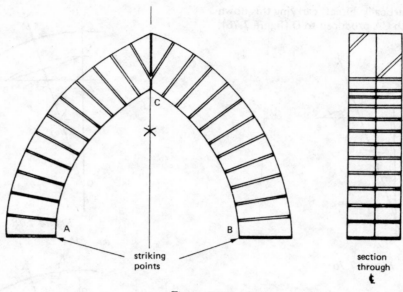

Figure 7.80

Since the Gothic arches are pointed, there will be no key brick.

Setting Out

(1) Draw the springing line AB and vertically bisect.
(2) Open the trammel heads a distance equal to AB and with point on A and B in turn draw the intrados to meet the vertical bisection at C.
(3) Further open the trammel heads to give the face depth and draw the extrados.
(4) Open the dividers 75 mm and mark in the voussoirs from the topmost point on the extrados down to the springing line, adjusting if necessary.

The procedure from this point regarding setting out, traversing and building is as for the semicircular arch, except for the fact that there is no key brick. In this case the cut bricks against the point on the arch are placed last.

Building

(1) Set the centre in position with a nail in each striking point, fastening a short length of line to each of these nails.
(2) Erect the brickwork either side as far as possible, keeping this ahead of the arch construction.
(3) Mark the position of the voussoirs on the centre and turn the arch evenly from each side, placing the cut bricks in last.
(4) Do not strike the centre until the brickwork

around the arch has been built at least up to the haunches. Finally, point up the soffit.

Camber Arch (figure 7.57)

This is a one-centred arch, known also as the flat or Georgian arch. It is basically a horizontal band cut from the face of a semicircular arch (figure 7.55) and probably the most common form is that enclosed in an equilateral triangle (figure 7.81). The soffit of the

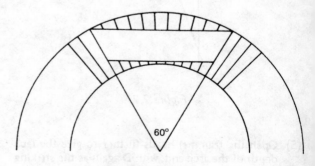

Figure 7.81

arch is not actually straight, but is given a slight camber of 1 mm in 100 mm to dispel the illusion of sagging. It has no great strength and therefore the maximum span should be kept to 1.350 mm.

Setting Out

(1) Draw the springing line and, at right-angles to this, the centre line (figure 7.82).

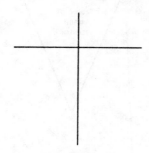

Figure 7.82

(2) Mark the span and draw the extrados 3 or 4 courses above the springing line as required (figure 7.83).

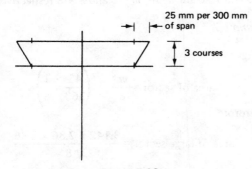

Figure 7.83

(3) Measure back the amount of skewback on the extrados above the squarehead. This is usually 25 mm per 300 mm of span and the skewback may now be drawn. Alternatively the angle of skewback is kept to 60° as previously explained (figure 7.83).

(4) Extend the skewbacks down to meet at the striking point and draw an arc from the top centre of the extrados down to each skewback (figure 7.84). This is known as the setting-out mark.

(5) The voussoirs are set out on this arc from the key brick down to the skewback and the joints are marked in by joining from these points down towards the striking point (figure 7.85).

(6) Draw the true line of the intrados by fixing nails at the springing points and on the centre to show the required amount of camber. Now spring a lath to touch all three nails, marking round the curve produced (figure 7.86).

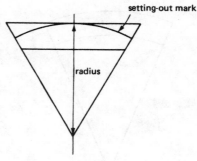

Figure 7.84

Figure 7.85

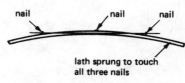

Figure 7.86

There are two methods of obtaining the templets for this arch. Method one is as follows. Complete the drawing of half the arch, including the key brick, on a thin sheet of ply and saw along the joint lines to obtain a templet for each brick. The templets are reversed to mark the voussoirs for the other half of the arch.

The traditional method is much more complicated and is considered beyond the scope of craft certificate students, as is the building of this arch.

AREAS OF ARCHES

Areas of Arch Faces

To find the area of an arch face it is necessary to find the area of the large sector and subtract from this the area of the small sector (figure 7.87). Thus a formula is required from which we can calculate the area of a sector. Consider figures 7.88, 7.89 and 7.90.

In figure 7.88, since the two radii shown are at right-angles to each other, the area of the small sector is ¼ the area of the circle (90/360 = ¼).

In figure 7.89, the radii are at 60° to each other and thus the area of the small sector is 60/360 or

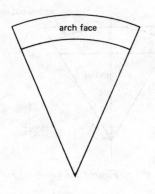

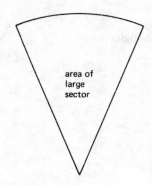

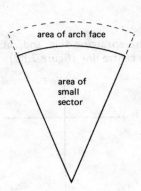

Figure 7.87

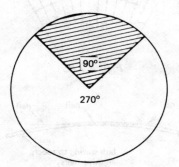

Figure 7.88

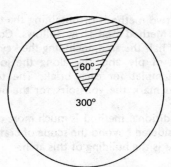

Figure 7.89

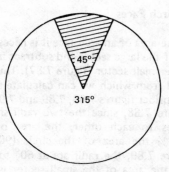

Figure 7.90

1/6 the area of the circle. Similarly, in figure 7.90, the area of the small sector is 45/360 or 1/8 the area of the circle. Once this is understood it follows that, since the formula for the area of a circle is πr^2, then the areas of the sectors shown in figures 7.88, 7.89 and 7.90 are $\pi r^2/4$, $\pi r^2/6$ and $\pi r^2/8$ respectively.

Example 7.7

In figure 7.91

$$\text{area of sector} = \frac{\pi r^2}{8} \quad \left(\frac{45}{360} = \frac{1}{8}\right)$$

Therefore

$$\text{area of large sector} = \frac{3.142 \times 2.86 \times 2.86}{8}$$

$$= 3.213 \text{ m}^2$$

$$\text{area of small sector} = \frac{3.142 \times 2.64 \times 2.64}{8}$$

$$= 2.737 \text{ m}^2$$

No.	Log
3.142	0.4972
2.86	0.4564
2.86	0.4564
	1.4100
8	0.9031
3213	0.5069
3.142	0.4972
2.64	0.4216
2.64	0.4216
	1.3404
8	0.9031
2737	0.4373

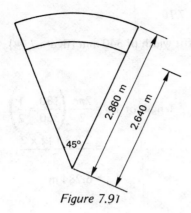

Figure 7.91

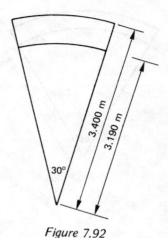

Figure 7.92

Therefore

$$\text{area of arch face} = 3.213 - 2.737$$

$$= 0.476 \text{ m}^2$$

Example 7.8

In figure 7.92

$$\text{area of sector} = \frac{\pi r^2}{12} \quad \left(\frac{30}{360} = \frac{1}{12}\right)$$

Therefore

$$\text{area of large sector} = \frac{3.142 \times 3.4 \times 3.4}{12}$$

$$= 3.027 \text{ m}^2$$

$$\text{area of small sector} = \frac{3.142 \times 3.19 \times 3.19}{12}$$

$$= 2.665 \text{ m}^2$$

No.	Log
3.142	0.4972
3.4	0.5315
3.4	0.5315
	1.5602
12	1.0792
3027	0.4810
3.142	0.4972
3.19	0.5038
3.19	0.5038
	1.5048
12	1.0792
2665	0.4256

Therefore

$$\text{area of arch face} = 3.027 - 2.665$$

$$= 0.362 \text{ m}^2$$

Areas of Arch Soffits

As previously explained, the soffit is the under-surface of the arch and it is necessary to understand how this area is calculated. As with areas of arch faces the angle at the centre is important since it determines what fraction the length of arc is of the total circumference. Once the length of the arc has been calculated this is multiplied by the soffit width to give the area.

Example 7.9

For a soffit width of 220 mm, in figure 7.93, since

$$\text{circumference of circle} = 2\pi r$$

and

$$\text{length of arc} = \frac{2\pi r}{6} \quad \left(\frac{60}{360} = \frac{1}{6}\right)$$

$$\text{length of arc} = \frac{2 \times 3.142 \times 3}{6}$$

$$= \frac{18.852}{6}$$

$$= 3.142 \text{ m}$$

Since

$$\text{soffit width} = 220 \text{ mm}$$

Then

$$\text{soffit area} = 3.142 \times 0.22$$

$$= 0.691 \text{ m}^2$$

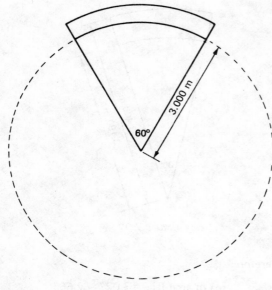

Figure 7.93

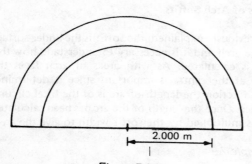

Figure 7.94

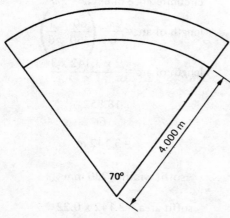

Figure 7.95

Example 7.10

For a soffit width of 340 mm (figure 7.94)

$$\text{length of arc} = \frac{2.\pi r}{2} \left(\frac{180}{360} = \frac{1}{2} \right)$$

$$= \frac{2 \times 3.142 \times 2}{2}$$

$$= 6.284 \text{ m}$$

Since

$$\text{soffit width} = 340 \text{ mm}$$

$$\text{area of soffit} = 6.284 \times 0.34$$

$$= 2.137 \text{ m}^2$$

If an angle were given at the centre which would not cancel conveniently with 360° the procedure would be as follows (figure 7.95).

For a soffit width of 210 mm (figure 7.95)

$$\text{length of arc} = \frac{2\pi r \times 70}{360}$$

$$= \frac{2 \times 3.142 \times 4 \times 70}{360}$$

$$= \frac{2 \times 3.142 \times 4 \times 7}{36}$$

$$= 4.888 \text{ m}$$

(cancelling zeros but cancel further if preferred)

No.	Log
2	0.3010
3.142	0.4972
4	0.6021
7	0.8451
	2.2454
36	1.5563
4888	0.6891

Since

$$\text{soffit width} = 210 \text{ mm}$$

$$\text{area of soffit} = 4.888 \times 0.21$$

$$= 1.026 \text{ m}^2$$

8
PIERS

TYPES OF PIER

Piers may be in two forms: (1) isolated (figure 8.1) or (2) attached (figure 8.2). The isolated pier is usually built to carry direct imposed loads and its maximum height must not exceed eight times its least dimension. The pier may be any shape, for example, square, rectangular, octagonal, etc., but if circular it is termed a column. Attached piers may occur on one or both sides of a wall, or be formed at the end of a wall. The purpose of the attached pier is to

(1) provide the wall with additional strength to counter lateral pressure
(2) accept indirect loading and transfer this load to the walling.

Walls with attached piers may be built in English, Flemish or garden wall bond, in all cases complying strictly with the rules of bonding. Tie-ing in walls to piers (figure 8.2) or vice versa (figure 8.3) must be not less than 56 mm and all cut bricks should be as large as possible.

The Building Regulations require external walls for small buildings, that is, garages, outbuildings, etc. to be supported by piers at the ends and also at intermediate lengths not exceeding 3.0 m.

Piers should receive the same protection as walls, and therefore caps are formed at the terminal height.

The minimum dimension for an attached pier is 200 mm square.

Note

A pier becomes a wall when the length exceeds four times the thickness.

Figure 8.5 illustrates ramp walling used to distribute lateral thrust and tie in raking courses.

A method of building attached piers using temporary aids is illustrated in figure 8.6.

COPINGS AND CAPS

A coping is the cover for the top of a free-standing wall. Its purpose is to prevent the penetration of water and frost into the topmost courses of the walling. Copings also stabilise the wall and provide an attractive finish for the termination of the work. When considering the type of coping required for a wall it is essential to consider the following factors

(1) the thickness of the wall
(2) cost considerations
(3) materials that are available
(4) the type of finish required.

To be effective a coping should

(1) be impervious to weather penetration
(2) contain the minimum number of mortar joints
(3) have adequate amount of fall on its topmost surface
(4) be provided with sufficient projection to allow water to fall clear of the wall face.

Materials used for copings are bricks, concrete, stone, slate, plastic and metal.

Types of Coping

Brick on Edge Coping

This should consist of dense pressed bricks bedded in cement mortar and laid with a fall of 6 mm per brick thickness of walling. When used on walls over one brick in thickness the coping should be bonded (figure 8.7).

Brick on Edge with Tile Creasing Course

The tile creasing course should consist of two courses of clay roofing tiles, bonded with a lap of at least 75 mm and bedded in cement mortar; the creasing course should project at least 38 mm on each side of the wall. The tile nibs can be used to form a decorative feature for the coping and the brick on edge should be bedded on a full mortar bed placed on the creasing course. When completed a cement mortar fillet is placed along the top of the projecting portion of the creasing course; this will assist in preventing water from entering the bed joint between bricks and tiles (figure 8.7).

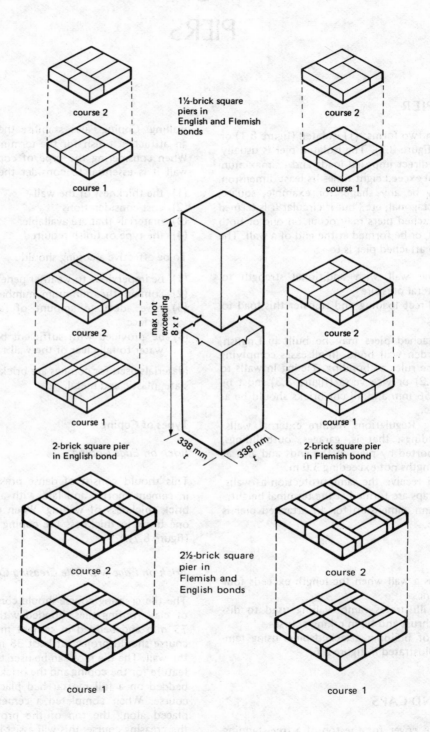

Figure 8.1 Isolated piers

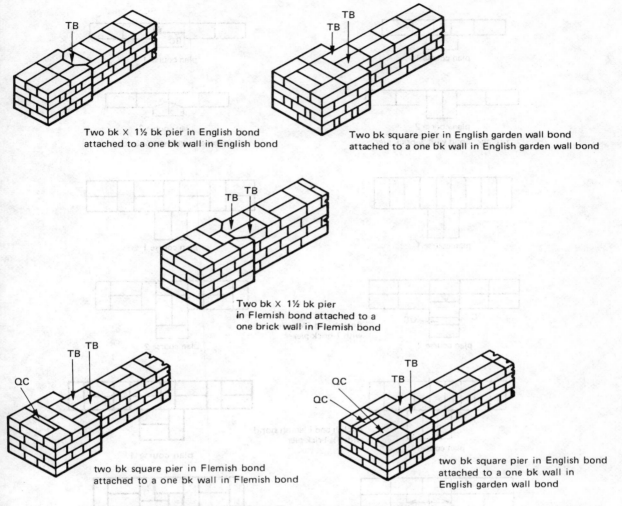

Two bk X 1½ bk pier in English bond
attached to a one bk wall in English bond

Two bk square pier in English garden wall bond
attached to a one bk wall in English garden wall bond

Two bk X 1½ bk pier
in Flemish bond attached to a
one brick wall in Flemish bond

two bk square pier in Flemish bond
attached to a one bk wall in Flemish bond

two bk square pier in English bond
attached to a one bk wall in
English garden wall bond

Figure 8.2   End piers in English, Flemish and garden wall bonds

Brick Saddle-back Coping

This coping is formed by cutting and bonding the bricks; it is extremely costly in labour and often in materials. When used, this type of coping should always be provided with a brick oversailing course and a cement mortar fillet. When fixing the coping it is essential to use three lines, that is, two side lines and one central top line, otherwise true alignment cannot be obtained (figure 8.7).

Radius or Circular Brick Coping

This type of coping is formed using engineering radius bricks. As with all brick copings it should be bedded

in cement mortar and provided with an oversailing course or a tile creasing course, which will protect the face of the wall from water (figure 8.7).

Concrete and Stone Copings

These copings can take varied shapes and when used should be fixed on full cement mortar beds. The advantage of these copings is that they contain only a minimum of mortar joints, which reduces the area of rain and frost penetration. Another factor is that they can be stabilised by using rebated joints, cramps and mortices. It is usual for these copings to be provided with a drip or throating, which allows water to fall clear of the wall face, as for feather-edge and plain or parallel copings (figure 8.7).

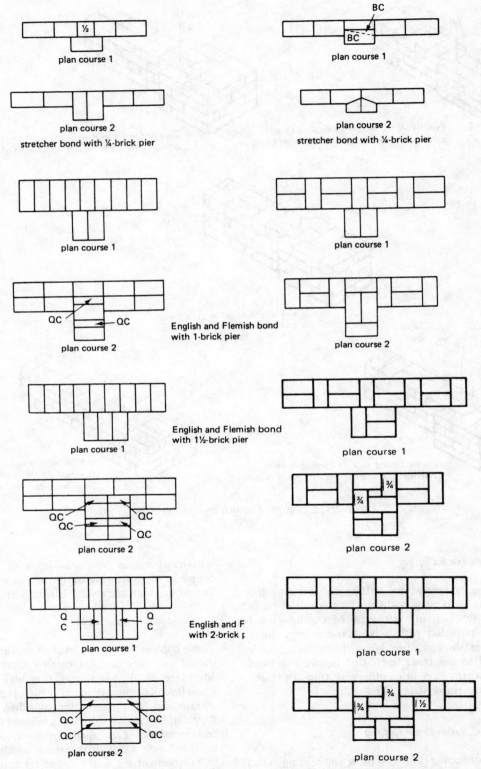

Figure 8.3 Single attached piers in English and Flemish bonds

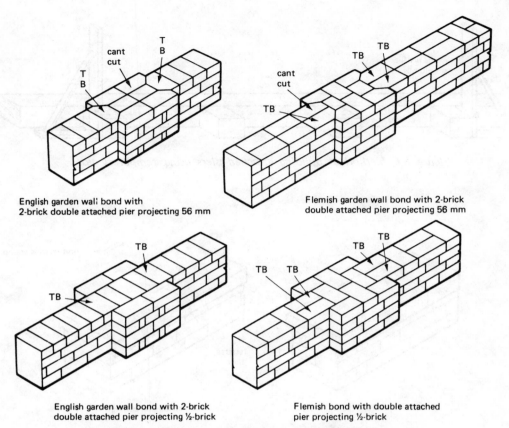

English garden wall bond with
2-brick double attached pier projecting 56 mm

Flemish garden wall bond with 2-brick
double attached pier projecting 56 mm

English garden wall bond with 2-brick
double attached pier projecting ½-brick

Flemish bond with double attached
pier projecting ½-brick

Figure 8.4 Double attached piers with English and Flemish bonds

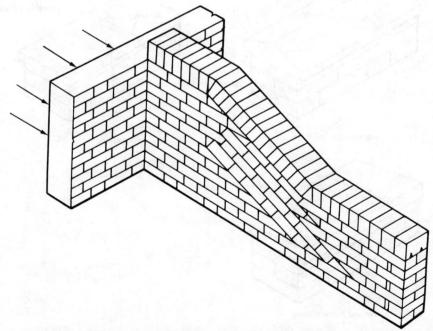

Figure 8.5 Ramp walling used to distribute lateral thrust and tie in raking courses

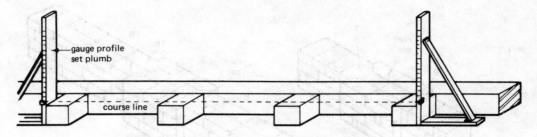

Figure 8.6 Method of building attached piers using temporary aids

gauge profile
set plumb

course line

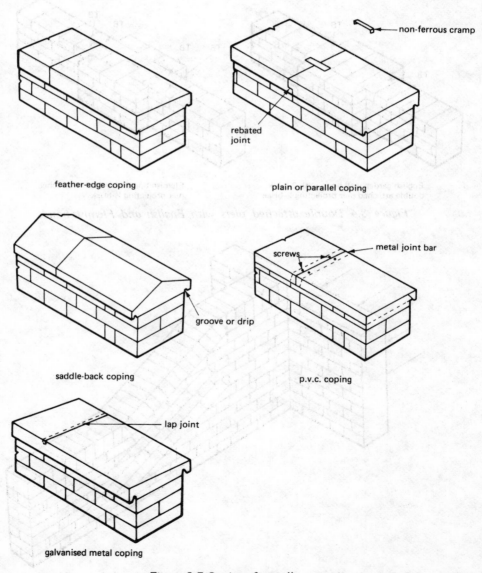

feather-edge coping

non-ferrous cramp

rebated joint

plain or parallel coping

saddle-back coping

groove or drip

p.v.c. coping

screws

metal joint bar

lap joint

galvanised metal coping

Figure 8.7 Copings for walls

Slate Copings

The shape of slate copings is limited to plain rectangular or feather edge (figure 8.7) because of the cost of the material, and the labour which would be involved if other shapes were required. Bedding is with a semi-stiff cement mortar; cramps or dowels should be used to provide additional stability.

Metal and Plastic Copings

These do not require any form of mortar bed; they are usually secured with the use of metal straps, lapped joints, and screws: see p.v.c. and galvanised metal copings (figure 8.7). Drips or throats are not provided. The metal coping should be galvanised or receive treatment to combat corrosion. These copings are becoming more common because of

(1) the reduced cost of the materials
(2) low fixing costs

Caps

These are used to provide the cover for attached and isolated piers since with copings it is advisable to provide a generous overhang, adequate fall and a minimum number of mortar joints (figure 8.8). Although

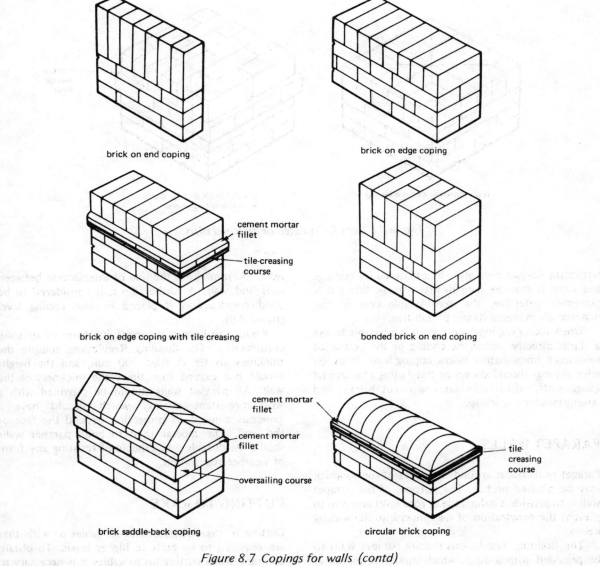

brick on end coping

brick on edge coping

cement mortar fillet

tile-creasing course

brick on edge coping with tile creasing

bonded brick on end coping

cement mortar fillet

cement mortar fillet

oversailing course

brick saddle-back coping

tile-creasing course

circular brick coping

Figure 8.7 Copings for walls (contd)

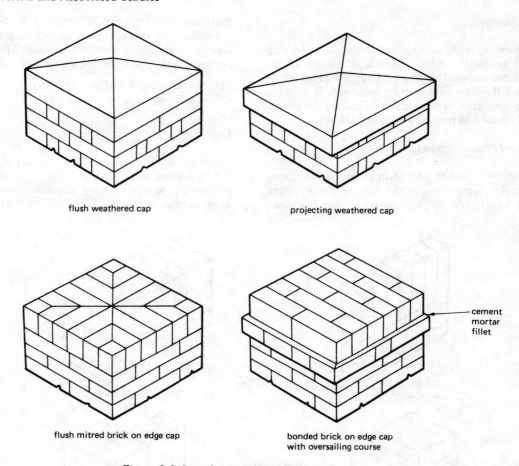

flush weathered cap

projecting weathered cap

flush mitred brick on edge cap

bonded brick on edge cap
with oversailing course

cement
mortar
fillet

Figure 8.8 Caps for isolated or attached piers

terracotta has not been given as a material for copings and caps, it may be seen in many areas. While it is extremely effective, the considerable cost of the material often makes its use prohibitive.

When fixing copings and caps it is advisable to use a d.p.c. directly below the coping or the course of brickwork immediately below coping level. Brick on edge copings should always be fixed using a horizontal gauge staff, which eliminates any cut bricks, and determines joint thickness.

PARAPET WALLS

Parapet walls occur around the edges of roofs, which may be pitched or flat. The purpose of the parapet wall is to provide a balustrade at roof level and also to prevent the penetration of dampness into the walling below.

The Building Regulations require parapet walls to be provided with a d.p.c. which should be placed at least 150 mm above the line of intersection between roof and wall. A further d.p.c. is considered to be good construction if placed beneath coping level (figure 8.9).

Parapet walls can be in cavity form or of solid construction. The Building Regulations require the thickness to be at least 200 mm, and the height should not exceed four times the thickness of the wall. All parapet walls should be provided with a weather-resistant coping, which should have a generous amount of projection beyond the face of the wall. The mortar used to build parapet walls should be durable and capable of resisting any form of weather penetration.

CUTTING TO RAKE

Cutting to rake often occurs on gables or walls that are required to be built to higher levels. To obtain accuracy when cutting up to gables it is necessary to

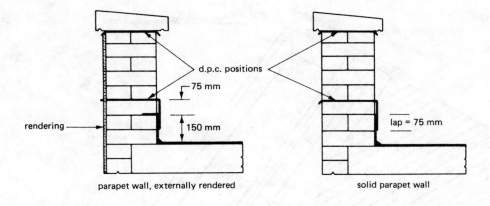

d.p.c. positions

75 mm

rendering

150 mm

parapet wall, externally rendered

lap = 75 mm

solid parapet wall

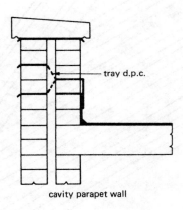

tray d.p.c.

cavity parapet wall

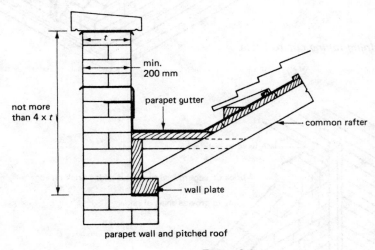

t

min. 200 mm

parapet gutter

not more than 4 x *t*

common rafter

wall plate

parapet wall and pitched roof

Figure 8.9

use pattern rafters, laths or battens, also profiles and dead men. Lines are required to provide the angle of rake; these are attached to laths nailed to the pattern rafters, or the pattern rafters themselves may be used to provide the angle of rake for the internal cavity walls of gables. To provide alignment for the gable, it is necessary to erect profiles wherever possible, but if this method cannot be adopted the use of a dead man will provide horizontal alignment (figure 8.10).

When cutting low-level walls to rake, the use of

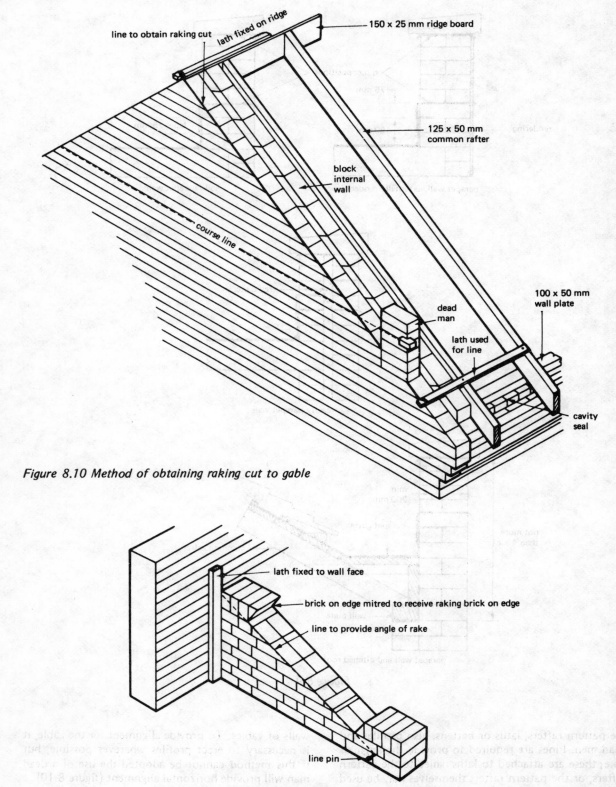

Figure 8.10 Method of obtaining raking cut to gable

Figure 8.11 Method of cutting to rake

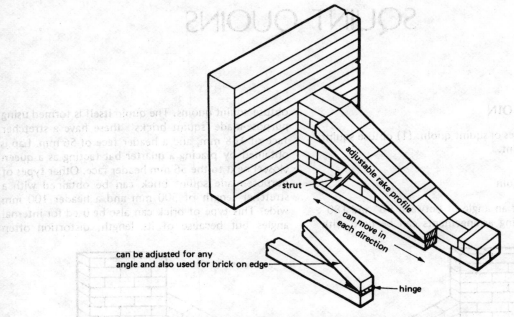

adjustable rake profile

strut

can move in each direction

can be adjusted for any angle and also used for brick on edge

hinge

Figure 8.12 Another method of cutting to rake

laths and lines is a very common method (figure 8.11). They provide the angle of rake, but are required to be altered to another position if brick on edge or other work is also to be constructed on the rake. Another method that is often more effective is the use of the adjustable raking profile. It can be moved horizontally and to any angle of rake, is easily adjusted and not affected by wind or weather (figure 8.12)

9
SQUINT QUOINS

TYPES OF QUOIN

There are two types of squint quoin: (1) obtuse squint and (2) acute squint.

Obtuse Squint Quoin

This is formed at an angle of between 90° and 180°. All normal bonding arrangements can be used with obtuse squint quoins. The quoin itself is formed using purpose-made 'squint bricks': these have a stretcher face of 168 mm, and a header face of 56 mm. Lap is obtained by placing a quarter bat (acting as a queen closer) next to the 56 mm header face. Other types of purpose-made squint brick can be obtained with a stretcher length of 300 mm and a header 100 mm wide. This type of brick can also be used for internal angles but because of its length, distortion often

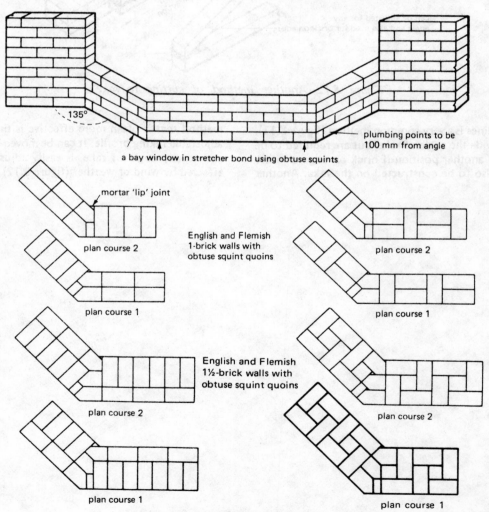

135°

plumbing points to be 100 mm from angle

a bay window in stretcher bond using obtuse squints

mortar 'lip' joint

plan course 2

plan course 1

English and Flemish 1-brick walls with obtuse squint quoins

plan course 2

plan course 1

plan course 2

plan course 1

English and Flemish 1½-brick walls with obtuse squint quoins

plan course 2

plan course 1

Figure 9.1 135° obtuse squints

occurs on both stretcher and header faces and the purpose-made squint with a 168 mm stretcher face is often preferred. When building with purpose-made squint bricks it is considered good practice to place the plumbing points at least 100 mm from the arris of the quoin, thus preventing difficulties which would arise from badly formed purpose-made bricks (figures 9.1 and 9.2).

Acute Squint Quoin

This squint quoin is used for angles less than 90°; normally 60° or 70° are the angles selected. The length of the stretcher face is normal brick length, but the header face will vary in length according to the angle required; the more acute the angle the greater the length of the header face (figure 9.3).

Recognising this fact, it will be appreciated that the normal queen closer cannot be used next to the quoin header, therefore a slip bat or triangular bat is placed next to the header to obtain quarter-brick lap. Normal bonding arrangements can be used with this type of squint angle; again plumbing should be at least 100 mm from the quoin arris (figure 9.3).

BONDS FOR CURVED WALLS

The type of bond used for building circular brickwork is normally determined by the radius of the curve. While header bond is very often used for this type of work it is possible to use English and Flemish bonds provided that the radius is suitable, or cutting the bricks is permissible (figures 9.4 and 9.5).

When the radius is less than 1.0 m, wide cross

Figure 9.2

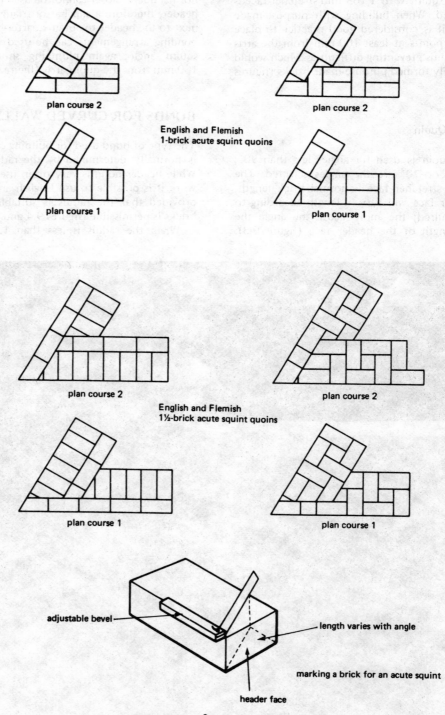

plan course 2

plan course 2

English and Flemish
1-brick acute squint quoins

plan course 1

plan course 1

plan course 2

plan course 2

English and Flemish
1½-brick acute squint quoins

plan course 1

plan course 1

adjustable bevel

length varies with angle

marking a brick for an acute squint

header face

Figure 9.3 60° acute squints

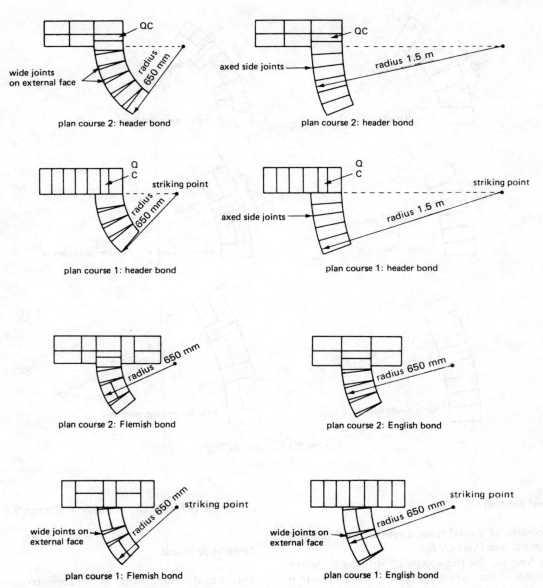

Figure 9.4 Bonding radius junctions

joints are formed on the convex face even if header bond is used and the bricks are uncut (figures 9.4 and 9.5). The external joints are also exceedingly wide if English or Flemish bonds are required and the bricks must be axed (figure 9.4). To use English or Flemish bonds without axing the bricks requires a radius of 2.0 m minimum (figure 9.5). When using header, English or Flemish bonds with radii between 1 and 2 m the use of half bats to form snap headers will considerably assist the bonding and reduce axed work on the stretchers.

ERECTING CURVED WALLS

When the craftsman is required to build walls that are circular in plan he must select the method and the aids that will be necessary to provide accuracy in construction. There are three methods of erecting curved walls; these are

(1) the trammel method
(2) templet or mould
(3) lead bricks and solid templets.

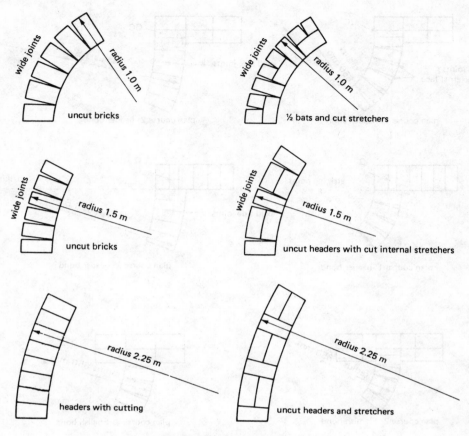

Figure 9.5 Curved walling

Trammel Method

This consists of a solid base, a vertical steel rod and the trammel lath (figure 9.6).

The base for the rod should be sufficiently heavy and capable of supporting the rod in a vertical position and so base sections of concrete are very adaptable. The height of the base can be increased as necessary to provide the stability that 'increased rod height' requires. The rod should be of mild steel with a diameter of 20 mm, but this may vary according to the length of rod required.

The trammel lath is formed of 62 x 12 mm timber which should have a pointed end to obtain the accuracy necessary in construction. Near the opposite end of the rod there is a hole formed for the rod, with a tolerance for the free movement of the trammel lath not exceeding 1.5 mm, otherwise accuracy would be difficult to maintain. The trammel lath should always be level during building. This is achieved by fixing two spring clips below the lath, which

can be moved vertically as required (figures 9.6—9.8).

Templet or Mould

This is made to the shape of the curve required, and consists of a framed structure which is placed on each course of brickwork as the work proceeds. To provide accuracy during construction the brickwork face of the straight walling must be completely aligned and plumb. The templet ends are projected beyond the curve enabling the templet to be accurately fixed in a plumbed position for checking of each course (figures 9.9 and 9.10).

Lead Bricks and Solid Templets

A solid timber templet is formed to the exact shape of the curve which may be concave or convex, according to the curved face required (figure 9.11).

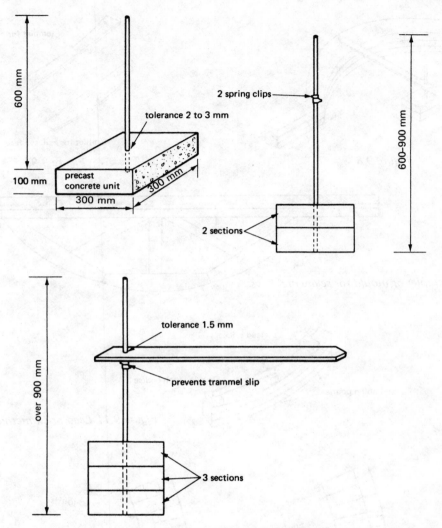

Figure 9.6 Fixing trammels and rods

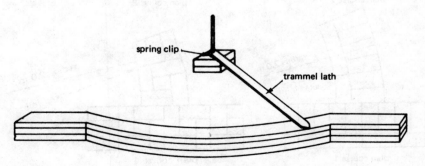

Figure 9.7 Use of trammel for curved work

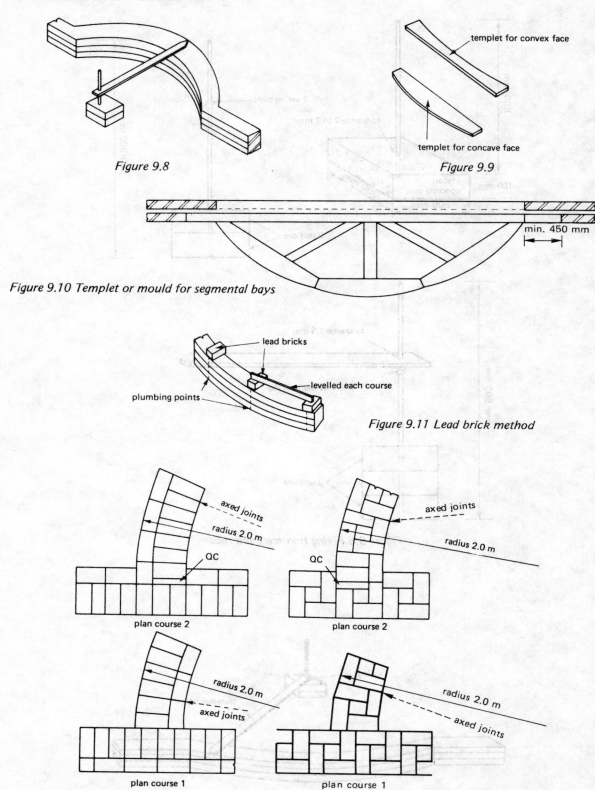

Figure 9.8

templet for convex face

templet for concave face

Figure 9.9

min. 450 mm

Figure 9.10 Templet or mould for segmental bays

lead bricks

levelled each course

plumbing points

Figure 9.11 Lead brick method

axed joints

radius 2.0 m

QC

plan course 2

axed joints

radius 2.0 m

QC

plan course 2

radius 2.0 m

axed joints

plan course 1

radius 2.0 m

axed joints

plan course 1

Figure 9.12 1½-brick junction walls in English and Flemish bond

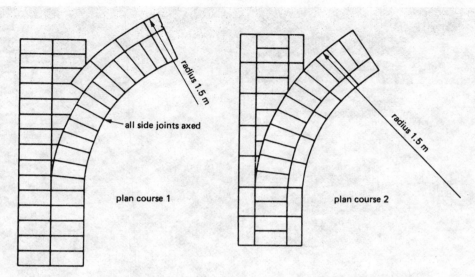

Figure 9.13 Junction of straight and circular walls in English bond

When a course of brickwork has been completed lead bricks are then bedded at fixed positions approximately 600 mm apart. These bricks are checked for gauge, levelled, and plumbed.

When all lead bricks have been satisfactorily set and checked, the bricks between each pair of lead bricks can then be laid, checked for level and positioned using the face templet with the lead bricks.

Junctions of Straight and Circular Walls

These are illustrated in figures 9.12—9.14.

FORMING CIRCULAR RAMPS

Circular ramps are often required at the ends of free-standing walls (figure 9.15). The ramp can be formed using two methods

(1) the horizontal trammel, or vertical profile trammel
(2) fixed templets

The trammel method using a horizontal beam is often used to form quadrant ramps. Although quite accurate,

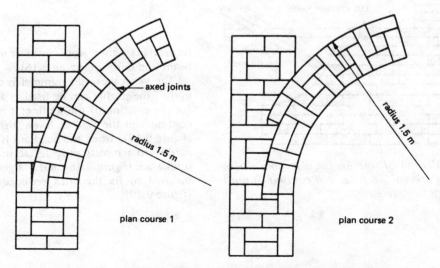

Figure 9.14 Junction of straight and circular walls in Flemish bond

Figure 9.15

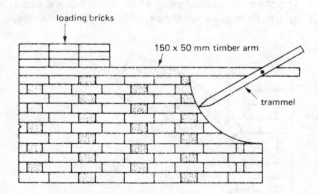

Figure 9.16 Method of forming circular ramp where loading bricks prevent work on wall continuing until circular ramp is completed

it prevents other work from continuing while the beam is in position (figure 9.16).

The vertical-profile trammel is considerably more useful: the profile can be used as a temporary quoin for the erection of the walling; it also allows work to continue on the wall at levels higher than the ramp (figure 9.17). When the quadrant is convex the use of hardboard templets provides accuracy in cutting the brickwork (figure 9.18) and a larger templet can then be used to fix the brick on edge around the ramp (figure 9.19).

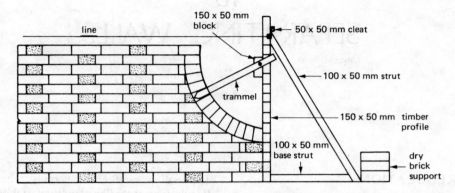

Figure 9.17 Method of forming circular ramp where the method does not prevent work from continuing

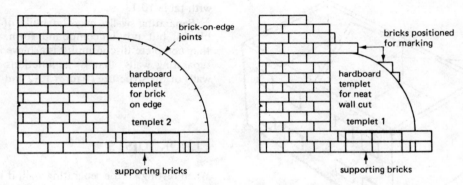

Figure 9.18 Using quadrant templets to form ramp

10
SEPARATING WALLS

BUILDING REGULATIONS

A separating, party or dividing wall is a wall constructed between two adjoining buildings, that is, between semi-detached houses or between each unit and the next in a row of terraced houses (figure 10.1).

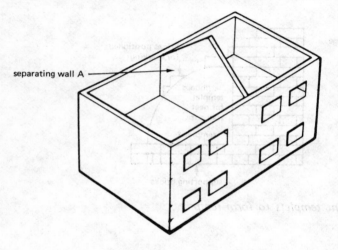

separating wall A

Figure 10.1 Separating wall between two houses

The wall is required to comply with the Building Regulations regarding structural stability, fire resistance and sound insulation. Separating walls may be of solid or cavity construction. The minimum thickness should be not less than 190 mm if solid, and 250 mm if formed with a cavity.

Separating walls should be built in compliance with table 10.1.

Separating walls may be built of brickwork or blocks, but not in compound form. It is essential that complete discontinuity is achieved at the ends of separating walls, that is, at junctions with the main walls of the structure (figures 10.2 and 10.3).

FLOOR JOISTS

When resting on the separating wall, it is essential that the joists do not pass beyond 100 mm of the wall thickness, and whenever possible the joists should be staggered, with a minimum of 100 mm between joists on one leaf and the joist supported by the other leaf (figure 10.4).

Table 10.1

Height of wall	Length of wall	Minimum thickness
Less than 3.5 m	Not exceeding 12 m	190 mm for whole height
Between 3.5 and 9 m	Not exceeding 9 m	190 mm for whole height
	Exceeding 9 m but not exceeding 12 m	290 mm for 1st storey height and 190 mm above

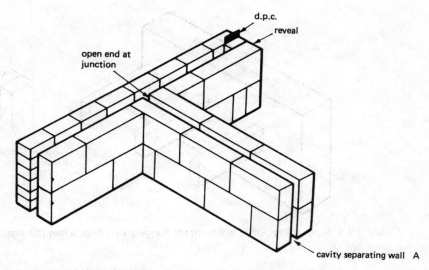

Figure 10.2 Junction of main walls and separating wall A

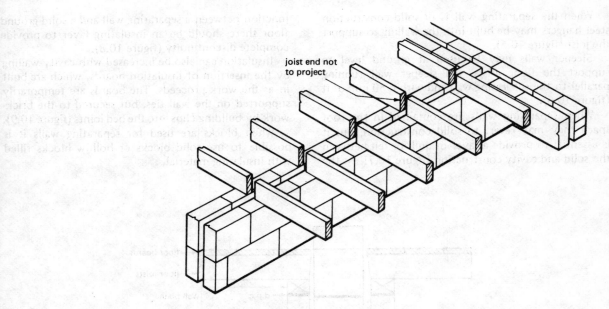

Figure 10.3 Joists set in line on cavity separating wall

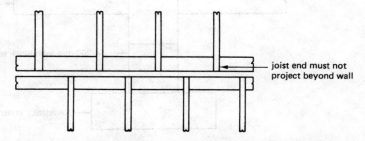

Figure 10.4 Staggered joists on cavity separating wall

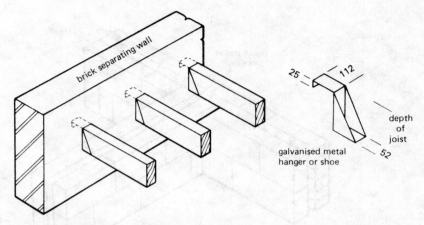

Figure 10.5 Supporting joists with metal hangers or shoes when the separating wall is solid

When the separating wall is of solid construction steel hangers may be built into the walling to support the joist (figure 10.5).

Sleeper walls may be used at ground level to support the floor joists, the sleeper wall running parallel to the separating wall and within 50 mm of it (figure 10.6).

When separating walls are contained in the roof space they may revert to solid construction, but it is essential to provide a layer of quilt or felt between the solid and cavity construction (figure 10.7). At the junction between a separating wall and a solid ground floor there should be an insulating layer to provide complete discontinuity (figure 10.8).

Insulation can also be increased with cavity walling by the insertion of insulation boards, which are built in as the work proceeds. The boards are temporarily supported on the wall ties, but secured to the brickwork by building clips into the bed joints (figure 10.9).

When blocks are used for separating walls it is possible to use solid blocks or hollow blocks filled with insulating material.

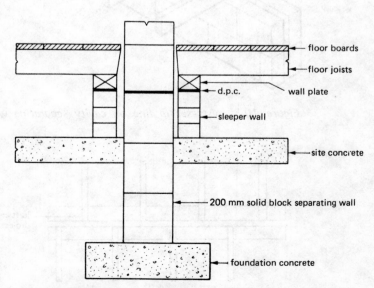

Figure 10.6 Solid separating wall with hollow timber floor

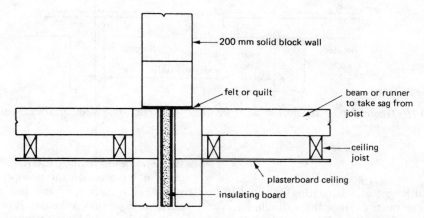

Figure 10.7 Method of reducing separating wall thickness; also from cavity to solid within the roof space

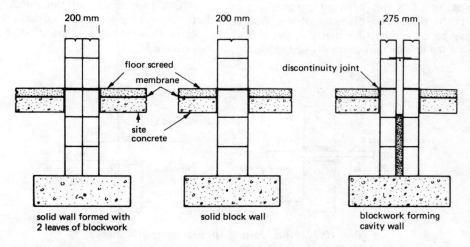

Figure 10.8 Blockwork separating walls supported on concrete strip foundations

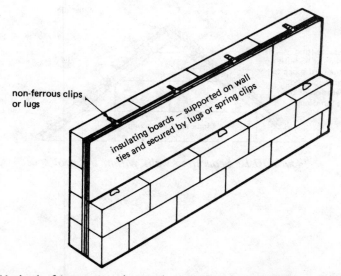

Figure 10.9 Method of improving the insulation of a cavity separating wall built with blocks

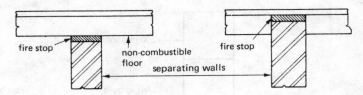

Figure 10.10 Treatment of separating walls at floor level (Building Regulations E5 (a) and (b))

FIRE STOPS

At the terminal height of a separating wall that is formed below a combustible floor there should be a fire stop, placed between wall and floor (figure 10.10).

Pipes and ducts may pass through separating walls, but there must be a fire stop between the pipe and the walling. If the pipe has a diameter of less than 25 mm it may be of combustible material, but if it exceeds 25 mm it must be non-combustible and have a diameter not exceeding 150 mm. Separating walls in houses not more than three storeys in height should have a fire resistance of not less than ½ hour, or 1 hour if four storeys in height. It is therefore essential to use the types of building material that will satisfy this requirement (figure 10.11).

It is a requirement of the Building Regulations that a separating wall must be built up in the roof space to ensure complete separation of the two buildings (figure 10.12).

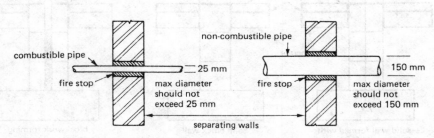

Figure 10.11 Pipes passing through separating walls

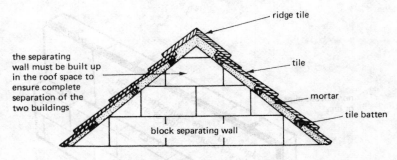

Figure 10.12 Separating walls within the roof space

11
FIREPLACES, FLUES, STACKS AND APPLIANCES

BUILDING REGULATIONS

Craft Certificate students are mainly concerned with solid fuel appliances as defined in the Building Regulations, and to a much lesser extent, gas appliances. The ensuing constructional requirements therefore refer to solid fuel appliances except where otherwise stated.

The chimney breast is built on a concrete foundation, which is widened sufficiently to give the required

spread as detailed in the Building Regulations. The foundation brickwork is usually erected to d.p.c. height and the oversite concrete is laid to a minimum thickness of 100 mm, off which the fender wall is built, also to d.p.c. height. This is a one-brick-thick wall and its dual purpose is to support the outer edges of the constructional hearth and the ends of the ground floor joists (figure 11.1). The area inside the fender wall and the fireplace recess is filled with hardcore and the constructional hearth is cast *in situ*,

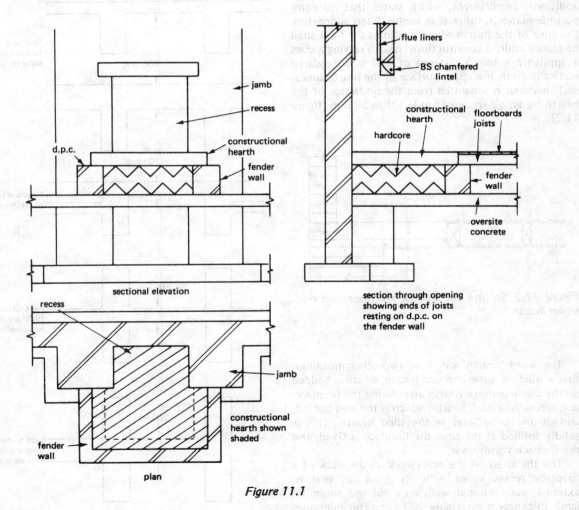

Figure 11.1

97

with its top surface at the same level as that of the finished floor. The minimum dimensions of the constructional hearth are stipulated in the Building Regulations (figure 11.1).

Regulations concerning constructional hearths are

(1) The minimum thickness permitted is 125 mm.
(2) The hearth must extend to the back of the recess.
(3) There must be a projection of at least 500 mm in front of the jambs.
(4) The hearth must extend at least 150 mm each side of the fireplace opening.
(5) The hearth surface must not be below the floor surface if the floor incorporates combustible material.
(6) If no recess exists the hearth is to contain a square measuring at least 840 x 840 mm.

Upper floor hearths are similarly governed with an additional requirement, which states that no combustible material, other than timber fillets supporting the edge of the hearth where it adjoins the floor shall be placed under a constructional hearth serving a class 1 appliance within a distance of 250 mm measured vertically from the upper surface of the hearth unless such material is separated from the underside of the hearth by an air space of not less than 50 mm (figure 11.2).

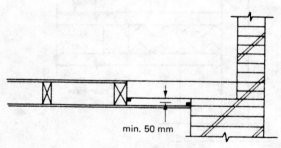

Figure 11.2 Section through upper floor constructional hearth

The word 'hearth' can have two other meanings: first a tiled, or superimposed hearth, which is bedded on the constructional hearth after fixing the fireplace; and secondly a back hearth, which is the area behind, and at the same level as the tiled hearth. This is solidly infilled at the time the fireplace is fixed, and the fireback stands on it.

The thickness of the brickwork at the back of a fireplace recess varies with its situation, that is, external wall, internal wall, etc., but the minimum jamb thickness is invariably 200 mm. This minimum is normally exceeded since the overall width of a tiled

surround is usually 1200 mm and, if minima were adhered to throughout, the width of the chimney breast would be only 980 mm (jambs 200 mm each and recess 580 mm) (figure 11.3).

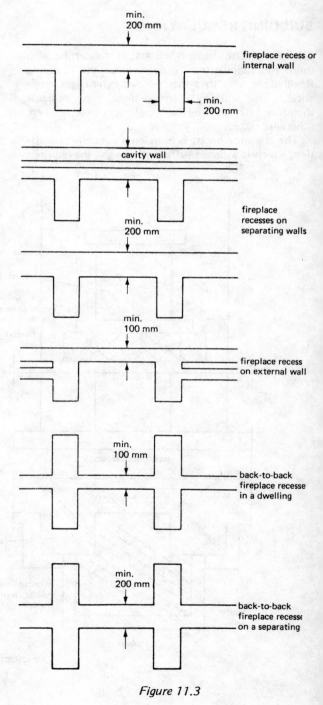

Figure 11.3

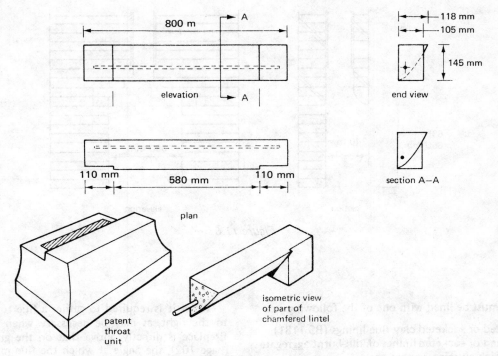

Figure 11.4 Fireplace lintel

CONSTRUCTION OF FIREPLACES AND FLUES

Fireplace Lintels

The height given is to the underside of the British Standard chamfered lintel (figure 11.6).

The BS chamfered lintel fulfils three functions, supporting the brickwork above the opening, forming the front of the throat and protecting the back of the tiled surround from the heat. If a chamfered lintel is not available when required, a reinforced lintel of rectangular section may be built in 975 mm above floor level, which will leave adequate room below for the insertion of a chamfered lintel when the fireplace is fixed. Other methods of supporting the brickwork include a rough, segmental arch or a suitable length of flat, mild-steel bar approximately 75 x 6 mm in section or a patent throat unit (figure 11.4), but a better method is to use a precast concrete raft lintel (figure 11.5).

The raft lintel should be built in about 800 mm above floor level. It carries the brickwork above the opening, and is holed at the centre to act as a starting point for the flue liners. It also eliminates the necessity for forming the gather at the throat, thus simplifying the work of the bricklayer.

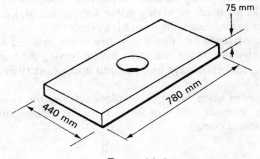

Figure 11.5

The dimensions for the builder's opening or fireplace recess depend on the appliance to be installed, but an opening 580 mm wide, 635–650 mm high from the floor level and 340 mm deep will accommodate almost any solid fuel domestic heating appliance (figure 11.6).

Where a raft lintel is not used, the builder's opening must be reduced in width to flue size as quickly as possible to support the flue liners. This is carried out by corbelling the brickwork towards the centre from each side until the width is 175–200 mm, where the first flue liner is rested. This can be accomplished in three courses (figure 11.6), and is known as *gathering over*.

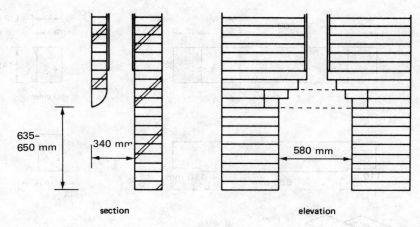

section   elevation

Figure 11.6

Flues must be lined with one of the following

(1) rebated or socketed clay flue linings (BS 1181)
(2) rebated or socketed linings of kiln-burnt aggregate and high-alumina cement or
(3) imperforate clay pipes with socketed joints (BS65: 1981).

The chimney may be constructed from concrete flue blocks made of, or having inside walls made of, kiln-burnt aggregate and high-alumina cement.

Joints between flue linings are either rebated or socketed and made in cement mortar leaving the bore clean, sockets always being placed uppermost (figure 11.7).

For an open fire in a fireplace setting, the minimum internal flue size is 185 mm if square and 200 mm if circular.

When it is required to gather a flue to the left or to the right, as may be necessary when an upstairs fireplace is directly above that on the ground floor (page 102), the angle at which the flue may be gathered over must never be less than 45° to the horizontal and is preferably kept to 60°. The bends shown in figure 11.8 are available for this purpose and these angles have proved adequate in all circumstances.

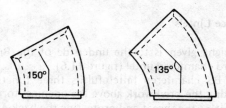

Figure 11.8

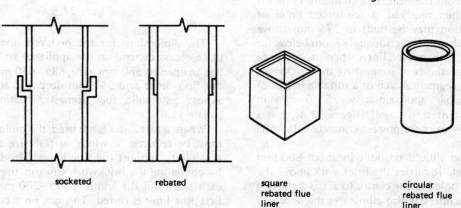

socketed   rebated   square   circular
                               rebated flue   rebated flue
                               liner   liner

Figure 11.7

The space within the chimney around the flue linings should be filled with weak mortar, but it is considered good practice to use vermiculite concrete, which will keep the flue warm and lessen the possibility of internal condensation, especially for chimneys on external walls. Dry vermiculite has been used for this purpose but this tends to pack down eventually, leaving the top of the flue liners in the chimney stack unprotected — where this protection is probably most important.

General fireplace construction is illustrated in figure 11.9.

Grouping the Flues

Above the ceiling joists the chimney breast must be reduced in width and this is carried out by arranging the flues to form the required stack size (figure 11.10).

Any flue in a chimney stack must be surrounded by and separated from any other flue by at least

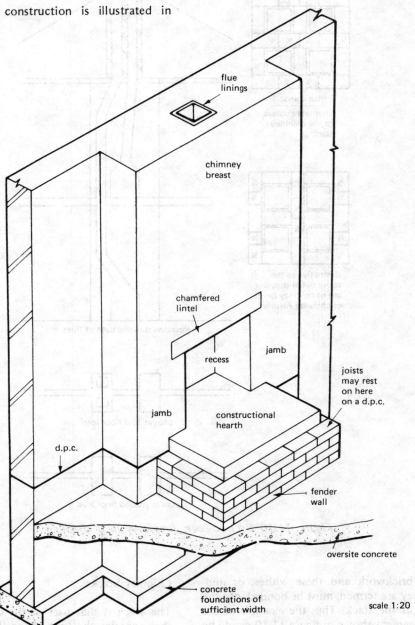

Figure 11.9 Fireplace construction

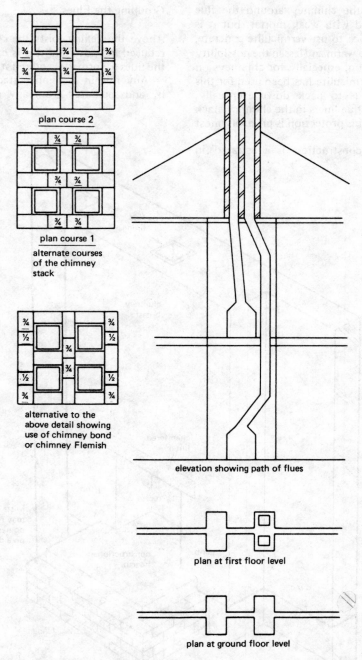

plan course 2

plan course 1
alternate courses
of the chimney
stack

alternative to the
above detail showing
use of chimney bond
or chimney Flemish

elevation showing path of flues

plan at first floor level

plan at ground floor level

Figure 11.10 Back-to-back fireplaces in a dwelling

100 mm of brickwork and these withes, or mid-feathers, as they are termed, must be bonded into the external wall of the stack. Thus the example of the bonding arrangement shown in figure 11.10 would be acceptable.

Chimney Stacks

The stack is that part of the chimney construction that contains the tops of the flues and which passes through and out of the roof. Thus the bricks used from this point will be facings rather than commons and the point of intersection between the stack and the roof is made weather-tight by the plumber at a later date, when he provides and fixes the lead flashings (see figure 11.11). Two courses above the lowest point of intersection between the stack and the roof a horizontal d.p.c. is built into the stack to arrest the downward passage of water (figure 11.11). While he is building the stack the bricklayer should rake out his bed joints 25 mm deep where required in preparation for the flashings, which should be pointed up in cement mortar on completion.

The terminal of a chimney stack refers not only to the pot but also to the cap, flaunching and any other constructional detail incorporated to complete the top of the stack.

Capping

It is preferable to complete the chimney stack with a precast concrete capping of sufficient width, and to

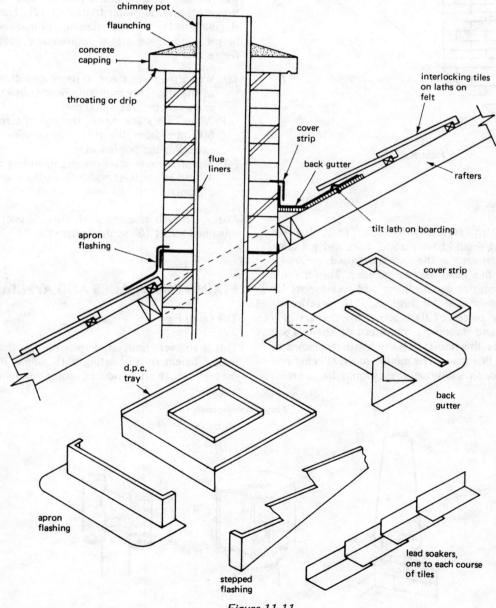

Figure 11.11

incorporate a drip or throating in order to shed rainwater clear of the stack below (see figure 11.11). Alternatively, oversailing courses can be built, keeping the maximum oversail to 38 mm for stability (see figure 11.12).

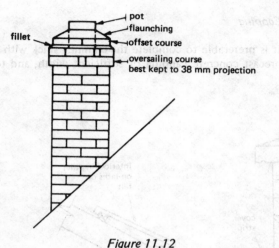

Figure 11.12

Chimney Pots

It is important, especially for open fires, that the chimney pot should have parallel sides and the same cross-sectional area as the flue liners used, otherwise the flow of flue gases may be impeded. Thus the rule of square pots for square liners and round pots for round liners should be kept to. The practice of placing four pieces of slate across the corners of a square flue and standing a round pot on these is bad practice since this results in four obstructive ledges to the flow of flue gases. The minimum overall chimney height for open solid-fuel fires should be approxi-

mately 5.5 m, but lesser heights may be suitable if a draught-inducing cowl as shown in figure 11.13 is fitted to increase the updraught.

Flaunching

This should be steeply sloping and formed in a strong mix of sand and cement, trowelled to a smooth finish. Its purpose is to retain the pot in position and to act as a weathering for the top of the stack. If an offset course is built on top of any oversails it should be weathered with a sand and cement fillet in order to throw off any rainwater (figure 11.12). The Building Regulations lay down minimum and maximum stack heights and these are as follows and as shown in figure 11.14.

(a) Where the stack passes through or within 600 mm of the ridge, the minimum height above the ridge excluding the pot is 600 mm.
(b) Where the stack passes through the roof above 600 mm from the ridge, the minimum height excluding the pot is 1 m.
(c) The maximum stack height including the pot is 4½ times the least width from the highest point of contact with the roof.

Note   In each case the pitch either side of the ridge must be at least 10° to the horizontal.

FIXING FIREPLACES AND APPLIANCES

The Open Fire

This is still very popular, despite the fact that it is the least efficient of all heating units, some 70—80 per cent of the heat produced going straight up the

draught-inducing cowls

Figure 11.13

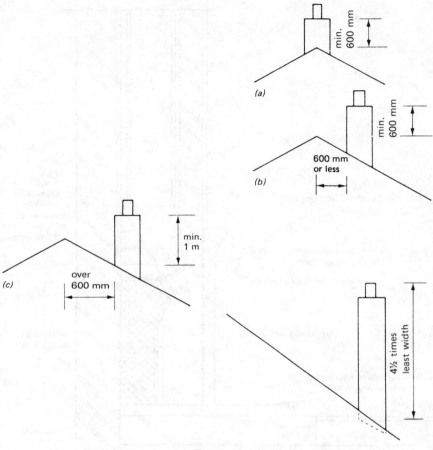

Figure 11.14

chimney. Fixing the fire is simple once the correct setting is there, but if the setting is incorrect or poorly constructed, problems such as draughty rooms and smoky fires are likely to arise once the fire is lit. The basic requirements for the open fire are

(1) the builder's opening (figure 11.6), including a suitable lintel (figure 11.4)
(2) the gatherings, to reduce the opening to flue size (figure 11.6)
(3) the throat, which must be carefully formed at the top of the opening along with the gatherings
(4) a suitable fireback and the infilling behind this
(5) a tiled or brick surround.

The opening and gatherings have already been described. We now consider the other requirements.

Throats (figure 11.15)

The front of the throat is formed by the chamfered lintel, either during construction or during fireplace fixing. Completion of throat formation is always carried out after building in the fireback, and many of the problems associated with open fires originate at this point.

The throat should be kept to 250 mm wide and 100 mm deep and must be carefully and smoothly formed to allow smoke to pass through without interruption. A slightly narrower throat than this would be satisfactory to remove the products of combustion, but would be too small for a brush to pass through for cleaning. If the throat is too large more air will pass through and may cause up to ten air changes per hour, which in turn will lead to excessive draughts within the room.

While the throat dimensions given will keep the number of air changes down to between four and six per hour, the insertion of an efficient throat restrictor (figure 11.16) at this point will bring this number down to 2½. A throat restrictor is usually fully opened until the fire is established, when it may be closed down to about 20 mm.

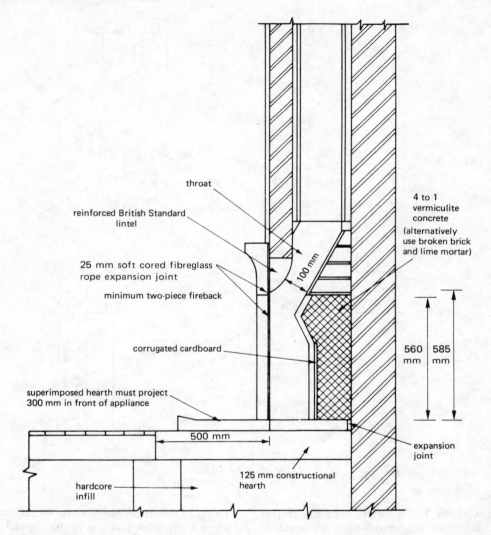

throat

reinforced British Standard
lintel

25 mm soft cored fibreglass
rope expansion joint

minimum two-piece fireback

corrugated cardboard

superimposed hearth must project
300 mm in front of appliance

hardcore
infill

100 mm

4 to 1
vermiculite
concrete
(alternatively
use broken brick
and lime mortar)

560
mm

585
mm

expansion
joint

500 mm

125 mm constructional
hearth

Figure 11.15 Fireplace installation to BS Code of Practice 403, using BS 1251: Part 2 Tiled surround

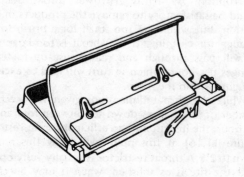

Figure 11.16

Firebacks (figures 11.17 and 11.18)

These are available in one, two, four and six pieces
and are manufactured in fireclay or concrete. Those
of fireclay tend to be costly in comparison with
concrete but are of better quality. It is preferable to
use at least a two piece since the lower half gets much
hotter and therefore expands more than the top half.
If a damaged fireback is being replaced it will be
easier to use a four or six piece.

One-piece Firebacks   While the one-piece fireback is
not included in BS 1251, it is available both in
concrete and fireclay, and is in fact probably the
most popular. The fireclay fireback has a groove

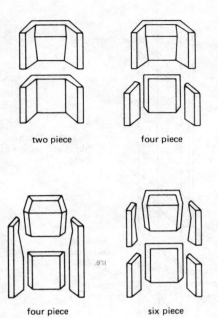

two piece   four piece

four piece   six piece

Figure 11.17 Firebacks complying with BS 1251: Part 1

incised where shown in figure 11.18 in order that it may be cut in two before fixing. The reason that it is not manufactured in two pieces is that it would distort during firing and the fit would be poor.

Surrounds

The surround provides an attractive finish and can be tiled or slabbed and fixed in one piece. Brick or stone surrounds are rapidly gaining popularity and these are built *in situ* by the craftsman.

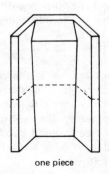

one piece

Figure 11.18

Fixing the Open Fire (figure 11.15)

(1) Knock away the wood blocks fastened to the base of the surround at each side. These are tacked on by the manufacturers so that the bottom tiles will not become chipped as the surround is being moved from place to place.

(2) Stand the surround against the chimney breast, making sure that it is central, and checking that the front and sides are vertical and the top is level. Prop safely in position.

(3) Carefully slide the tiled hearth into position, packing it up at the back as necessary to make a good fit with the surround (some hearths pass under the surround and need lifting up to meet the underside). Level the hearth in both directions, packing the front up as required. Packing should be carried out in soft materials since slate, etc. may chip the tiles around the sides. Small wooden wedges are ideal for this purpose.

(4) Form the back hearth preferably from a precast material such as part of a concrete paving slab. This should be bedded on stiff mortar and be large enough at least in area to support the fireback. Keep the slab 6–8 mm away from the tiled hearth so that an expansion joint can be made when the tiled hearth is finally fixed. Alternatively the hearth can be floated in concrete but this means waiting for this to harden before it will support the weight of the fireback.

(5) Mark the position of the fixing lugs so that pallets or plugs can be inserted in the correct position. It will help too if at this stage the surround is pencilled round on the brickwork as a guide for future alignment.

(6) Remove the tiled hearth and insert the bottom half of the fireback on to the back hearth. This should be placed to fit evenly 6–8 mm away from both sides of the surround to form yet another expansion joint.

(7) Remove the surround, place corrugated cardboard round the back of the fireback and carefully fill round with 4:1 vermiculite concrete. Take care not to move the fireback at this stage or the surround may not fit.

(8) Place a strip of fibreglass string on top of the fireback and stand the top section on this. It should be kept back about 2 mm rather than forward to keep heat away from this edge. Fill round this as before and brick up to the gatherings, forming the throat above the fireback as smoothly a possible.

(9) Make fixings in the position previously marked.

(10) Place a strip of fibreglass rope up one side of the fireback, across the front of the lintel and down

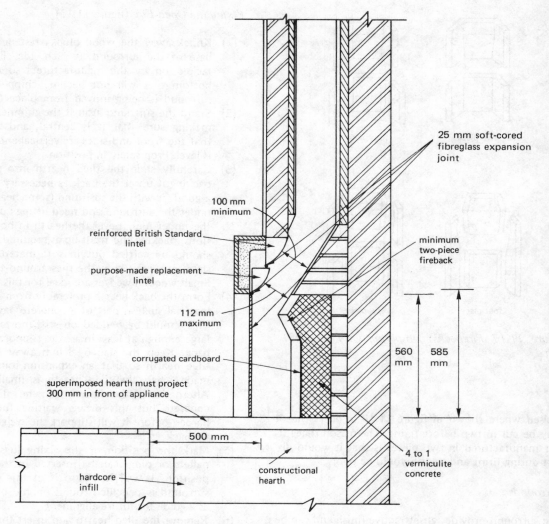

Figure 11.19 Fireplace installation using a 150 mm projecting surround

Labels on figure:

25 mm soft-cored fibreglass expansion joint

100 mm minimum

reinforced British Standard lintel

purpose-made replacement lintel

minimum two-piece fireback

112 mm maximum

corrugated cardboard

superimposed hearth must project 300 mm in front of appliance

560 mm

585 mm

500 mm

hardcore infill

constructional hearth

4 to 1 vermiculite concrete

---

the other side. Water glass will cause it to stick to this surface.

(11) Replace the tiled surround and screw through the lugs into the prepared fixings.

(12) Place a strip of fibreglass rope against the back hearth and solidly bed the tiled hearth on sand—lime mortar, sliding it into position to compress the fibreglass rope.

(13) Place the grate and fret and clean down the fireplace as necessary, allowing only a small fire for a day or so to allow the work to thoroughly dry out.

Projecting Tiled Surrounds (figure 11.19)

These are occasionally encountered and, apart from requiring an extra lintel, are fixed in exactly the same way as the flat-backed type already explained. The sidecheeks are set forward in the recess, leaving about 6 mm for the fibreglass rope between these and the tiled surround, and the purpose-made lintel is bedded flush with the front of the sidecheeks.

Inset Open Fires (figure 11.20)

This is sometimes known as the continuous or all-night burner and, while it is simple and relatively cheap to fix, it is not very efficient. It is fitted into a tiled surround up against the existing fireback which, incidentally, must be in good condition since more heat is produced than with an ordinary basket and fret. Fixing is as follows.

(1) If, on inspection, the existing fireback is unsound

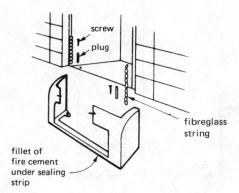

Figure 11.20

it should be removed and a new two, four, or six-piece fireback inserted.

(2) Place the complete fire in position, ascertaining that the firebars are approximately 5 mm away from the fireback since these may expand considerably when heated. Shorten the firebars if necessary.

(3) Replace the firefront and mark the hearth through the fixing lugs to prepare the position of the fixing plugs. Before removing, mark the edges of the firefront where they fit against the fireback.

(4) Remove the firefront, drill the hearth and fix metal or fibreglass plugging.

(5) Fix fibreglass string with water glass on the front edges of the fireback where it is marked, and, turning the firefront upside down, liberally smear the underside of the sealing strip with fire cement.

(6) Now place the firefront in position, compressing the fibreglass string, and screw firmly down into the prepared plugs.

(7) Clean off any excess fire cement and place the firebars and other fitments in position.

The air for combustion for the fires so far described is taken from the room, through the front and up through the firebars. This means that cold air must travel at low level in the room towards the fire creating draughts. The next fire overcomes this problem to some extent, by taking its air for combustion from under the floor. This is little problem with hollow timber floors but with solid floors a duct must be provided under the floor, usually with pipes led to airbricks, situated in external walls.

Deep Ashpit Fires (figure 11.21)

General fixing principles are as listed below, but the manufacturer's instructions should be studied and carefully followed.

(1) If it is known during construction that a deep ashpit fire is to be fitted, an opening should be left in the fender wall and part of the constructional hearth should be left out to accommodate the ashpit and air-supply pipe. If, on the other hand, the fire is a replacement, the back hearth must be excavated 350 mm deep for the deep ashpit and the excavation must be carried forward through the constructional hearth and fender wall so that the 75 mm-diameter rigid PVC pipe may connect to the nozzle on the ashpit and project to the underfloor area.

(2) The ashpit must be concreted in place to the correct height and must be perfectly level; the rigid PVC pipe is then fitted on to the nozzle and the brickwork to the fender wall made good.

(3) The air control is next provided for by a vertical damper rod which fastens to a circular butterfly valve within the nozzle. This is controlled at hearth level by a damper lever which operates through a metal tube.

(4) Bed the fire basket in fire cement to the finished hearth level and fill round this with a weak concrete mixture. Also make good the constructional hearth.

(5) Bed both sidecheeks and the bottom back section in mortar, prop firmly in position and fill round with broken brick and lime mortar. Fix the top back section in position and complete the infill, forming the throat as described previously.

(6) Bed the replacement lintel between 460 and 560 mm from floor level since the fire will be at hearth level and may tend to smoke if the opening is too high.

(7) Build or fix the surround, not forgetting the fibreglass rope up the sides and across the front of the fireback and lintel.

(8) Provide for expansion between the fire basket and the hearth and fix the hearth, first removing the damper rod from the tube.

(9) Place the front in position and position the damper rod through this back down the tube, tightening the butterfly damper on to the rod from the inside using the thumbscrew.

(10) Lastly, place the damper lever on top of the damper rod.

The fires mentioned so far heat the room by radiation only, any warm air produced being carried straight up the chimney with the smoke and fumes. A considerable amount of heat also passes through the fireback and infilling and is simply wasted, especially if the chimney breast is built on an external wall. The next fire is more efficient in that it intercepts some of this heat

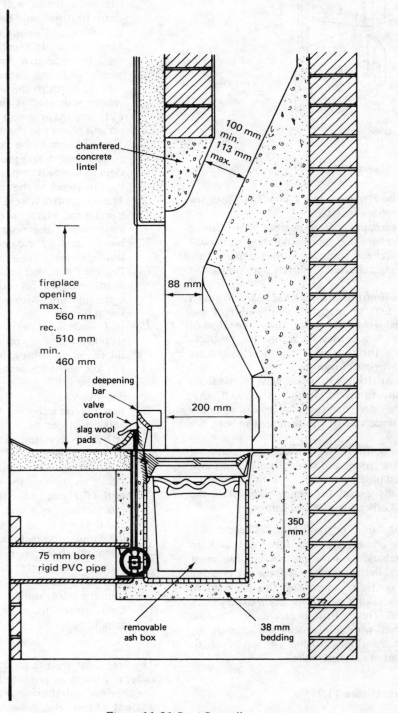

chamfered
concrete
lintel

100 mm
min.
113 mm
max.

88 mm

fireplace
opening
max.
560 mm
rec.
510 mm
min.
460 mm

200 mm

deepening
bar

valve
control

slag wool
pads

350
mm

75 mm bore
rigid PVC pipe

removable
ash box

38 mm
bedding

Figure 11.21 Baxi Burnall

lost through the fireback and transfers it back into the room as a current of warm air.

Inset Convector Fires (figure 11.22)

Essential points with the fixing of this fire are as follows

(1) The builder's opening must be of the correct size to form the convection chamber around the unit. The clearance at the back and sides of the unit must be kept to 50–60 mm since if it is much larger the air will not be warmed sufficiently and if it is any smaller the warm air supply may be severely restricted. The brickwork around the unit should be smooth, and preferably rendered, to assist the flow of warm air.
(2) An inlet must be provided for the air to get behind the unit. This is accomplished by the provision of two low-level inlet grilles, one either side of the surround, which are ducted into the convection chamber. The ducts will have to be cut through the jambs unless provision was originally made.
(3) There must be an exit for the hot air to get into the room and this is provided above the fireplace opening by a grille, which must be situated just below the raft lintel.
(4) The seal between the surround and the unit must be perfect or unwanted cold air will get into the convection chamber.
(5) A reinforced concrete flue connecting block must be cast to the manufacturer's specifications, with the hole of the correct diameter and in the right place, built in at the correct height. The joint between this and the flue outlet must be very carefully made using 25 mm-thick fibreglass rope and the clamping ring supplied (figure 11.23). Any leaks at this point could lose all the warmed air up the flue.
(6) When the surround is fixed against the unit, fibreglass rope is used in between the two, to seal the joints.

If required, the warmed air may be ducted to adjoining or upstairs rooms and a further improvement is to supply the inlet air from under the floor, which will cut down draughts in the room containing the fire.

Double-cased Convector Fires

Some manufacturers provide an integral convector jacket in cast or sheet iron. In this instance, any space between the casing and the brickwork should be filled with a non-combustible insulating material such as mineral wool. This will help prevent any heat losses through the brickwork.

Boiler Models

Both single and double-case convector fires are available with back boilers; these are supplied and usually already built into the appliance, together with the boiler flue, by the manufacturer. As with open fires, where the pipes pass through the brickwork in the jamb they should be passed through metal sleeves well plugged with fibreglass string to obviate the possibility of air leaks.

Inset Boiler Fires (figure 11.24)

The main points in the fixing of these fires are as follows.

(1) Sweep and check the flue.
(2) If necessary, form the correct size of builder's opening.
(3) Check the appliance for damage: this must be put right before installation.
(4) Set the back hearth, surround and tiled hearth.
(5) Using the templet provided, mark and drill out for the holding-down bolts.
(6) Using thin, greased steel skids, ease the appliance into position, sealing to the surround with fibreglass rope, and bolt to the hearth.
(7) Connect primaries and heating mains and test before infilling with vermiculite concrete.
(8) Extend the flue preferably into the first liner and infill around the extension with vermiculite concrete.

Freestanding Boilers (figure 11.25)

The bend from the appliance back to the chimney breast must be not less than 45° to the horizontal and must have access for cleaning. The flue pipe should be sleeved through the brickwork, caulked up with fibreglass rope and covered with a sealing plate to mask possible plaster cracks. No combustible material must be placed within 200 mm of the sides and below this pipe nor within 300 mm above. At the base of the flue a condensate trap and bowl must be provided and a double-sealed sootbox built in for cleaning if externally placed. Single sealed is suitable if internal.

Converting from Solid Fuel to Gas-burning Appliances (class 2)

When solid-fuel fires are removed and gas fires are installed the local gas board will supply specifications for preparation of the opening (figure 11.26). In each case illustrated the brickwork above the opening must be adequately supported and the flue must be smoothly gathered to the required dimensions (max 225 x

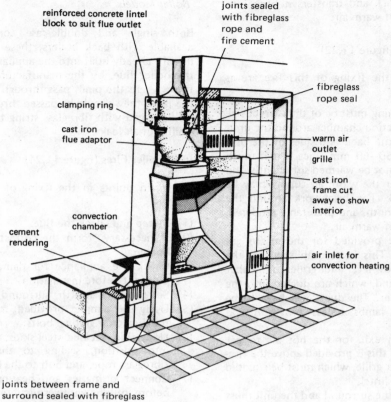

reinforced concrete lintel
block to suit flue outlet

joints sealed
with fibreglass
rope and
fire cement

fibreglass
rope seal

clamping ring

cast iron
flue adaptor

warm air
outlet
grille

cast iron
frame cut
away to show
interior

convection
chamber

cement
rendering

air inlet for
convection heating

joints between frame and
surround sealed with fibreglass
rope and fire cement

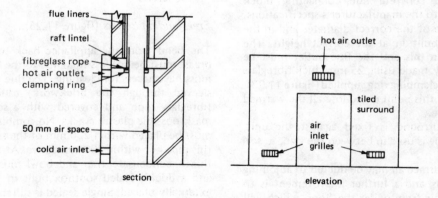

flue liners

raft lintel

fibreglass rope

hot air outlet

clamping ring

50 mm air space

cold air inlet

section

hot air outlet

tiled
surround

air
inlet
grilles

elevation

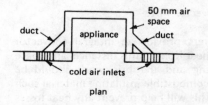

50 mm air
space

duct

appliance

duct

cold air inlets

plan

Figure 11.22 Single-cased convector fire

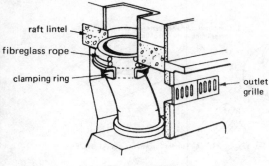

raft lintel

fibreglass rope

clamping ring

outlet grille

Figure 11.23

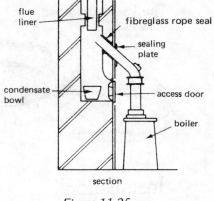

flue liner

fibreglass rope seal

sealing plate

condensate bowl

access door

boiler

section

Figure 11.25

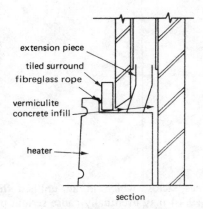

extension piece

tiled surround

fibreglass rope

vermiculite concrete infill

heater

section

Figure 11.24 Inset boiler heater

gas boilers are installed into properties built before 1966 it is vital to line the flue with a liner of the correct diameter, usually 100 or 125 mm depending on the size of the boiler.

This operation is carried out from the top of the stack. A flexible stainless steel flue lining such as a Kopex lining is used. This is very light and strong and is available on drums or in packs (figure 11.27). When one of these linings is to be installed the overall length required should be carefully measured, and, if in doubt, drop a length of line down the flue with a weight on the end. It is important to have sufficient length since making a joint is impossible, and yet to over-order is a costly business (the price varies according to the diameter). When this is to hand, installation is as follows.

(1) Remove flaunching and chimney pot from the chimney stack.
(2) Clean and core the existing flue as necessary.
(3) Drop the draw cord on the nose cone down the flue and insert the lining, at the same time gently

225 mm). The freestanding fire must have a hearth, minimum 750 x 225 x 60 mm, the top 13 mm of which must be non-combustible. This is not just a case of leaving a hole for the flue pipe, as is often commonly thought, but an opening must be formed to a specific height, width, and depth, depending on whether the fire is wall fixed or freestanding. When

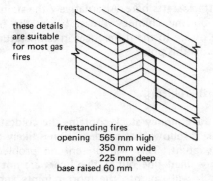

these details are suitable for most gas fires

freestanding fires
opening   565 mm high
           350 mm wide
           225 mm deep
base raised 60 mm

base of opening min. 200 mm from floor level

chase for gas supply

wall-fixed fires
opening   460 mm high
           350 mm wide
           225 mm deep

Figure 11.26

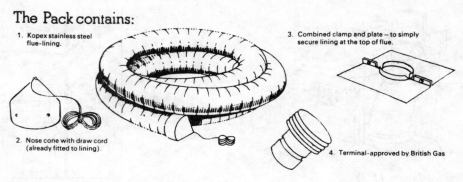

The Pack contains:

1. Kopex stainless steel flue-lining.

2. Nose cone with draw cord (already fitted to lining).

3. Combined clamp and plate – to simply secure lining at the top of flue.

4. Terminal-approved by British Gas

Installation

Gas Installations only suggested methods

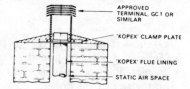

APPROVED TERMINAL, GC1 OR SIMILAR

'KOPEX' CLAMP PLATE

'KOPEX' FLUE LINING

STATIC AIR SPACE

Figure 11.27 The Kopex flue lining

pulling on the cord from below to help negotiate the bends.

(4) When the lining is in position, fasten the top end by means of the clamp plate which is flaunched over to the tube with mortar.

(5) Fix the terminal by pressing it into the mortar and add further flaunching as necessary.

Note

An insulating filling placed around the lining before fitting the clamp seal is considered good practice and will help prevent internal condensation.

CONDENSATION IN FLUES

Flue gases produced by all fuels contain varying amounts of water vapour, which will condense on the sides of the flue if the surface is cold enough. Most of the condensation occurs almost immediately after the fire is lit since this is the time when the flue is at its coldest. When the fire has been burning for a while the flue liners absorb heat as the flue gases pass upwards, and in a well-constructed flue the condensation should stop. If, however, the chimney is on an external wall or is poorly insulated a longer warming-up period will be required and thus more condensation will occur. The condensate contains, among other

substances, acids, which in an unlined flue (built before 1966) may eventually cause serious problems.

Internal Problems

Depending on the amount of condensation occurring within the flue, sooner or later the acids will penetrate the brickwork and plastering surrounding the flue, causing unsightly brown or black staining on decorations, the staining often being accompanied by a pungent odour. More often than not the stains occur just below ceiling level in the ground floor. This is where the first bend occurs in a chimney breast (see figure 11.10). The acids usually pass down the vertical section of the flue on to this sloping surface and the acid attack starts here. Alternatively the staining may occur on the chimney breast below the bedroom ceiling owing to the differences in temperature above and below this point.

External Problems

Since the chimney stack contains the coldest part of the flue it follows that condensation is likely to occur at this point. This will present no problems in a properly lined flue but if flue liners are not present the acids will eat into the mortar joints and in this case will react with the tricalcium aluminate (C_3A)

constituent of the cement in the mortar. The reaction between the two will cause a softening, expansion and deterioration of the joints, often resulting in a curved chimney stack and possible eventual collapse. This reaction is known as sulphate attack, which is more fully explained in Volume 1, chapter 3. Lined flues too are liable to sulphate attack, but from external sources. Industrial chimneys, road transport, etc. may all add carbon monoxide and sulphurous products to the atmosphere. Thus rainwater running down the sides of chimney stacks may be polluted with these substances and capillary action may occur, resulting in sulphate attack at the mortar joints.

Other possible causes of sulphate attack are the bricks used in the chimney construction. These may themselves contain sulphates which, when dissolved in rainwater, may migrate to the joints, once again causing expansion as the sulphates in solution react with the tricalcium aluminate constituent in the cement.

TRANSFER OF HEAT

It is necessary to have fires and other heating appliances in buildings in order to maintain a comfortable temperature for occupants. We must remember, however, that heat loss is inevitable since heat will always travel from a hot body to a cooler one and this transfer of heat occurs by conduction, convection, radiation or a combination of these.

Conduction

Heat transfer in solids always occurs by conduction and perhaps the simplest example of this is shown when one end of a poker is placed in a coal fire and the handle is touched at intervals. Whether or not the handle is shielded from the fire, it will soon become hot and the longer the poker is left in the fire the hotter it will become.

Similar experiments can be carried out by holding strips of various metals such as brass, tin, copper, etc. in the flame of a bunsen burner. In these cases too it will be found that heat travel will be fairly rapid and it may be concluded that metals are good conductors of heat.

Next try samples of brick, glass, concrete and similar building materials and it will be found that while heat is still transferred along the material it is a much slower process. Thus these are only fair conductors, perhaps better described as fair insulators. Water too is a fair insulator and can be tested by heating in a test tube, holding a bunsen burner as shown in figure 11.28. While the surface of the water can be observed to be boiling, the water at the bottom of the test tube remains cold for a considerable length of time.

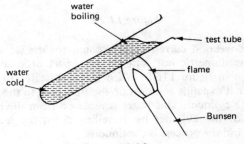

Figure 11.28

Good insulators usually contain small pockets of air and table 11.1 lists examples of conductors and insulators.

It will be noticed from table 11.1 that the denser materials are the best conductors of heat and the lighter the material the better the insulator.

Convection

This occurs in gases and liquids and is due to movement of the material. Thus the basic difference between conduction and convection is that there is no movement of the material during conduction.

When water is heated it expands, becomes less dense and rises. This is easily demonstrated by

Table 11.1

Good conductors	Fair conductors/insulators	Good insulators
Aluminium	Concrete	Glass fibre
Tin	Brick	Cork
Iron	Stone	Aerated concrete
Brass	Glass	Still air
Lead	Water	Polystyrene foam
Zinc	Sand/cement plaster	Exfoliated vermiculite
Copper	Sand/cement screeds	
	Plasterboard	Lightweight plaster

heating water at the bottom of a test tube, in which case all the water soon gets hot fairly quickly (figure 11.29). Compare this result with the experiment for conduction in water (figure 11.28).

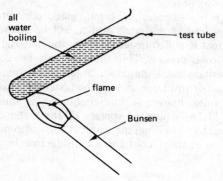

Figure 11.29

Convection currents also account for the working of the domestic hot water system, part of which is shown in figure 11.30. When water is heated in the boiler it expands and rises up the flow pipe to the top of the cylinder. This water is replaced from the bottom of the cylinder by travelling down the return pipe and the process is continuous.

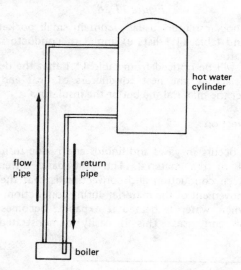

Figure 11.30

Convection in air is similar but more violent since air flows more easily and expands much more quickly than water. The open fire burns because of convection currents as shown in figure 11.31. As the fire burns, the air directly above in the flue expands, becomes less dense and rises up the flue, being replaced by further air from within the room, which in turn is replaced by air from under doors and around

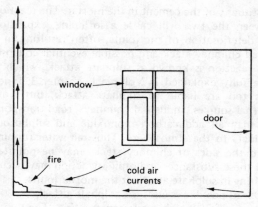

Figure 11.31

windows. If all air inlets to the room were effectively sealed with draught excluders, etc. the fire would cease to burn because of air starvation. On the other hand, if the fireplace throat is large and there are plenty of air inlets into the room, cold draughts will be created, which explains why someone sitting in a warm room with a huge fire may suffer from cold feet since any cold draughts travel at low level towards the fire.

The drawing of fresh air into a room is called ventilation and, as previously stated, 2½ air changes per hour are considered ideal. A lesser number than this causes the air in the room to become stale and a greater number causes draughts to occur and burns more fuel.

A simple experiment to demonstrate the necessity for air and the principle of convection is to construct a small airtight box having a removable glass front and two small tubes leading from it, as shown in

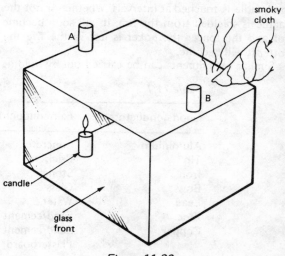

Figure 11.32

figure 11.32. Experiments can then be carried out as follows.

(1) To show convection currents in air: remove the glass front and place a lighted candle directly below tube A. Replace the front and hold a smoky cloth above tube B. It will be seen that smoke is drawn down tube B, across the box and rises up tube A. Thus cold air for combustion must be entering at tube B.

(2) To show that fire will not burn without air: set up the burning candle as before and seal the top of tube B. If the box is airtight the oxygen inside will soon be used up and, with no more able to enter, the flame will soon die out.

Radiation

A familiar sight in the building trade during winter is a brazier, which is not only used to warm cold hands, but also to thaw out frozen aggregates, sands, bricks and the water tub occasionally. Here is an excellent example of radiation, showing that heat can be transferred through the atmosphere either upwards, downwards or laterally (see figure 11.33). This cannot be

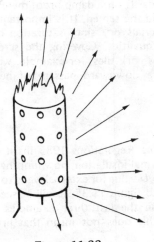

Figure 11.33

due to conduction, where heat is transferred through a solid and, as already explained, heat always rises because of convection. Radiant heat is transmitted in waves from a source to a recipient and if a barrier were placed between the two the heat would immediately be cut off. Perhaps the best example of radiation is the sun, which may be very hot one minute, but if a cloud were to blot it out the transfer of heat stops as quickly as switching off an electric light. The instant the cloud clears, the radiant heat is restored to its original intensity.

The open coal fire gives off heat by radiation. Once again it is noticeable that if the fire is obscured from a body by an object such as a piece of furniture all heat is immediately cut off.

THERMAL INSULATION

Heat escapes from buildings, not only through gaps around ill-fitting doors and windows but also through walls, floors and roofs owing to a combination of conduction, convection and radiation. In the past when fuel was plentiful, dwellings were kept warm by having very large fires, no thought being given to the escape of heat. The modern trend, however, is to have as small a fire as possible and to keep the heat produced inside the building, which saves on fuel and is thus much more economical. The thermal transmittance or U-value of the fabric of a building is a measure of its ability to conduct heat out of the building, and the greater the U-value the greater the heat loss. When heat passes through a cavity wall, for example, resistance to the passage of heat is offered by each surface or material (see figure 11.34).

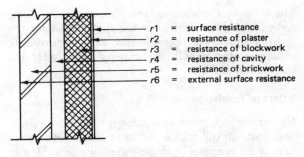

$r1$ = surface resistance
$r2$ = resistance of plaster
$r3$ = resistance of blockwork
$r4$ = resistance of cavity
$r5$ = resistance of brickwork
$r6$ = external surface resistance

Figure 11.34

The heat from within the building is said to have to 'penetrate' the internal wall surface, pass through the plaster and blockwork by conduction, cross the cavity by radiation, pass through the brickwork by conduction and 'leave' the external surface by a combination of radiation and convection. This overall resistance to the passage of heat is calculated and the thermal transmittance or U-value is the reciprocal of this. Modern buildings are required to be built to a certain standard to prevent excessive heat loss and the maximum permissible U-value for an external wall is laid down as 0.6 W/m² °C. While this figure will mean little to the student at this point, the examples of U-values for certain walls given in figure 11.35 should help.

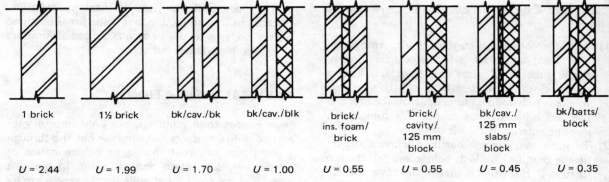

1 brick	1½ brick	bk/cav./bk	bk/cav./blk	brick/ ins. foam/ brick	brick/ cavity/ 125 mm block	bk/cav./ 125 mm slabs/ block	bk/batts/ block
U = 2.44	U = 1.99	U = 1.70	U = 1.00	U = 0.55	U = 0.55	U = 0.45	U = 0.35

(*U* = values in W/m²/°C, walls plastered one side, cavities unventilated)

Figure 11.35

The total heat loss through walling can be found by multiplying the appropriate *U*-value by the area of the wall and multiplying the result by the temperature difference between the inside and the outside.

From the examples shown, the poor *U*-value of a one-brick solid wall is yet another reason why this is unsuitable for the external wall of a dwelling.

Note

The *U*-values given are variable to some extent since, for example, the density of the bricks used for the facework may vary considerably and, as previously explained, the denser the material, the faster the conduction of heat.

Effect of Insulation

No matter how much insulation is built into a dwelling, heat will still escape, but at a considerably slower rate. For example, a single-glazed window (3 mm glass) has a *U*-value of approximately 5.68 W/m² °C, whereas double glazing will cut this by half to 2.84. To be effective the gap between the panes of glass should be at least 15 mm, and preferably 20 mm.

Floors

Heat will escape downwards through hollow timber ground floors by conduction and radiation and the inclusion of an insulating quilt or insulating boards between the joists and the floorboards will improve this (figure 11.36). The thermal resistance of solid floors too can be improved by placing insulating quilts or boards on a damp-proof membrane between the slab and the screed. It is important to keep insulating materials dry since saturation cuts down the insulating qualities. Covering the screed with such materials as cork tiles, for example, will also help to prevent the downward passage of heat (see figure 11.37).

Walls

The Building Regulations 1985 insist on high standards of thermal insulation, now limiting the maximum permissible *U*-value for external walls to 0.6 W/m²/°C. Referring to figure 11.35, it will be seen that for new construction the last three examples are eminently suitable. This does not mean that brick/insulation/

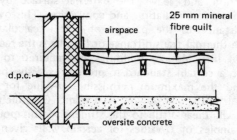

Figure 11.36

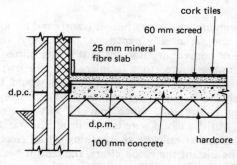

Figure 11.37

brick cannot be utilised (see above) but the use of urea formaldehyde foam is referred to in the Building Regulations, and not all situations are suitable for this injection method.

It is necessary to box in any air grates across the cavity so that air entering is directed below timber ground floors and not into the cavity, which is also sealed at the eaves (figure 11.38*d*).

If air movement occurred in cavities it would interfere with the insulating properties of the wall and the *U*-value would rise; heat would be lost by conduction through the inner leaf, radiation across the cavity and convection due to air movement.

Methods of sealing around the airbrick include the use of slate or purpose made ducts (figure 11.38*f*). In any event, the duct should slope outwards so that moisture is not directed towards the internal leaf, and if the duct is above the horizontal d.p.c. it should have a short length of d.p.c. tray above it sloping down towards the external leaf.

One of the simplest and most effective methods of bringing the *U*-value well below that required by the Building Regulations is to build into the cavity, panels of glass fibres known as Dritherm batts. These are produced by Messrs Pilkingtons of St Helens and, while unsuitable for completed buildings, they are ideal for new construction. The batts are fibreglass resin-bonded slabs 1200 x 450 mm in area and are either 50, 65 or 75 mm thick, depending on the width of the cavity. They are rot-proof, odourless and will not transmit water to the internal leaf. The cavity insulation should commence either from the top of the cavity infill, or from a row of wall ties placed one or two courses below the horizontal d.p.c., and at 600 mm c/c. This must be taken into consideration when building in the first row of wall ties above d.p.c. (now spaced at the usual 900 mm c/c). The external leaf is built up to just above this level and the first course of insulation should be carefully cut, either with a sharp knife or with a bricklayer's trowel, and fitted into place. The d.p.c. is now bedded on the inner leaf, and blockwork is brought up to the same height as the insulation (figure 11.38*a*, *b* and *c*).

From this point, either leaf may be built up just above the next row of wall ties which should be 450 mm above the first. After clearing away any surplus mortar from the back of the walling, the next course of insulation is fitted into place and the other leaf is again raised to the same height. It is important that mortar does not fall on top of the insulation, and a suitable batten should be provided with this in mind.

The use of this form of insulation will improve the *U*-value by an average of 65 per cent, depending on the other materials used for the wall and the thickness of the batts. For example, a wall built of brick

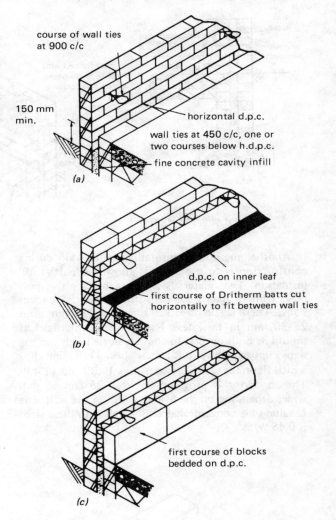

Figure 11.38a, b, c

external leaf, 75 mm Dritherm batts and 100 mm aerated concrete block has a *U*-value of approximately 0.35 W/m^2 °C.

When building in Dritherm batts *DO NOT*:

1. Build both leaves first and then push the insulation down the cavity.
2. Roughly tear the insulation, for example where extra wall ties are used (every 300 mm max. vertically at an unbonded jamb). Make any cuts carefully in order not to leave any gaps.
3. Leave any gaps in the insulation. All joints must be closely butted and free from mortar.
4. Use the Dritherm on edge, for example where small pieces are required. The face, not the edge, must be adjacent to the wall surface.

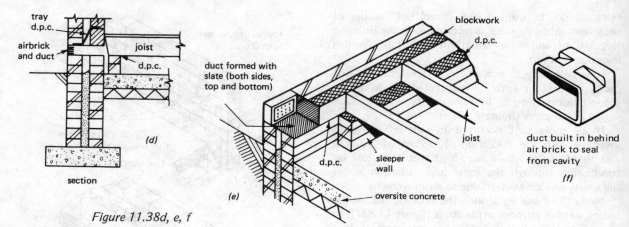

Figure 11.38d, e, f

Another method of insulating cavity walls during construction is to use what is known as 'partial fill' insulation. Two materials are available for this purpose, namely glass wool slabs and polyurethane foam slabs. These again are 1200 mm x 450 mm, but 25–30 mm in thickness. Normally the internal leaf should be built up first to above wall tie level, and the slabs clipped tightly to it by the special retaining clips which fit on the wall ties (see figures 11.39 and 11.40). Thus a cavity is provided between 25 and 50 mm wide, depending on the overall width of the wall. The *U*-value of a conventional cavity wall with these slabs is 0.45 W/m² °C.

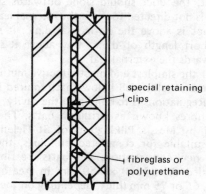

Figure 11.40

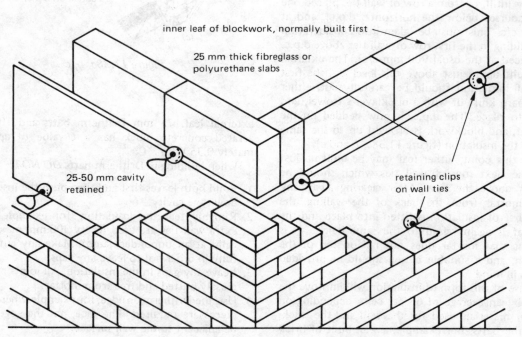

Figure 11.39

There is a further method of achieving the mandatory 0.6 W/m²/°C. A wall consisting of brick/cavity/125 mm aerated lightweight concrete block, with lightweight plaster on the internal face, will satisfy the Building Regulations.

The minimum thickness of separating walls given in the Building Regulations 1985 is 200 mm solid, which will conduct heat to a certain extent, depending on the materials used for construction and the temperature difference existing either side. Thus, because of the still-air cavity, it is preferable for thermal insulation purposes to have cavity walls in this situation.

Ceilings

Plasterboard can at best only be described as a fair insulator and the transmittance of heat through ceilings is relatively fast (see figure 11.41), being aided by the fact that any heat produced in the room below will rise towards this point because of convection currents. Thus it is essential that insulation is incorporated here and this is usually accomplished by filling in on top of the plasterboard between the ceiling joists with glass fibre, exfoliated vermiculite, expanded polystyrene, etc. If this is done, however, it is essential to leave out the insulation below a cold

water storage cistern or this may freeze up during cold spells. It is considered good practice to insulate around and over the cistern in order to maintain the temperature of the water contained at above freezing point.

Roofs

Insulation can be included in pitched roofs as well as between the ceiling joists, for example, by placing a mineral-fibre quilt over the rafters before the sarking felt is applied. Other methods include the use of insulating boards, cork board, wood wool slabs, etc.

SOUND INSULATION

Sound is a form of energy: it is measured in decibels (dB), and its transmission may be airborne, impact or due to a combination of the two. Part E of the Building Regulations is concerned with sound insulation and states that separating walls and floors separating dwellings (flats) must provide adequate resistance to the passage of sound. Methods of achieving the necessary sound insulation value are outlined in part E.

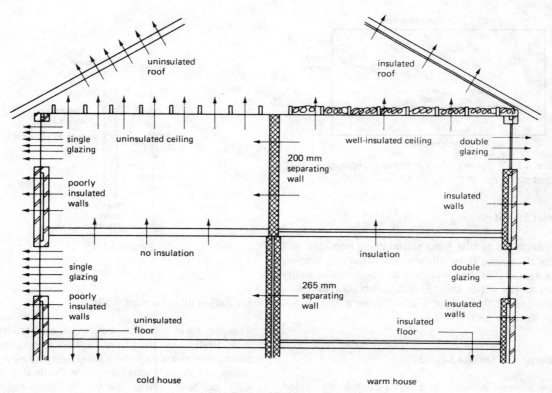

Figure 11.41

Airborne Sound

Airborne sound may be direct or reflected: it passes through the atmosphere from the source of the sound to the ear. If the surface of the wall, floor or ceiling is hard and smooth, for example, glazed tiles, dense concrete, etc., the reflection may be considerable (see figure 11.42), whereas if the surfaces are soft and porous, for example, thick carpets, heavy curtains, polystyrene tiles, etc., the sound will be absorbed and reflection will be much less (see figure 11.43). Good examples of airborne sounds are those from aeroplanes, vehicles, radios, voices, etc.

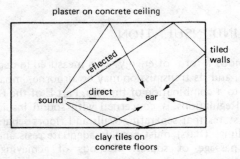

Figure 11.42 Sound absorption and reflection: hard, smooth surfaces

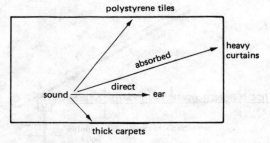

Figure 11.43 Sound absorption and reflection: soft, porous surfaces

Impact Sound

Impact sound results from sound being imparted to a material causing it to vibrate. For example, tapping on a radiator at one end of a building can be easily heard at the other end. Footsteps on staircases too, depending on the surface finish, may be plainly heard some distance away.

Airborne and Impact Sound

When a noise is made in a room completely sealed off from another and yet still heard on the other side, this is a combination of airborne and impact sound. A simple example of this is a neighbour's radio or television in one of a pair of semi-detached houses. The noise passes through the air from the source to the wall, causing it to vibrate and transmit some of the noise to the occupants in the adjoining building. Depending on the constructional materials, some of the sound would be reflected, some absorbed and some transmitted (see figures 11.44 and 11.45).

Since the wall in figure 11.45 is thicker and heavier, less vibration will take place, resulting in less sound transmission.

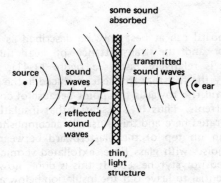

Figure 11.44

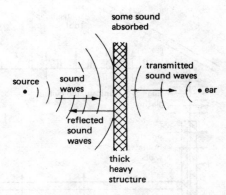

Figure 11.45

Insulation of Separating Walls

This can be effected by reducing the vibration of the wall, that is, by thickening and increasing the mass per square metre or, alternatively, including a still-air cavity, which is a good insulator. Walls that comply with the requirements of the Building Regulations 1985 are shown in figure 11.46.

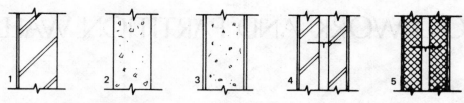

Figure 11.46 *Separating walls complying with the Building Regulations for sound insulation*

(1) bricks or blocks plastered on at least one face
(2) dense *in situ* concrete or concrete panels with grouted joints
(3) lightweight concrete with plaster on both faces
(4) brick or block skins with both faces plastered, and butterfly wall ties; minimum 50 mm cavity with the average mass/m² not less than 415 kg
(5) two lightweight concrete leaves with a minimum 75 mm cavity, plastered both sides, with the average mass/m² not less than 250 kg.

In cases 1–3 the average mass/m² must be not less than 415 kg.

Insulation of Floors

Increasing the mass per square metre is one method (with plaster, concrete or screed) but, where timber floors are used, pockets of still air coupled with light, porous materials will satisfy the Building Regulations (figure 11.47).

Insulation of Windows

Double or triple glazing is to be preferred, with as large an air space as possible, 75–150 mm, between the sheets of glass.

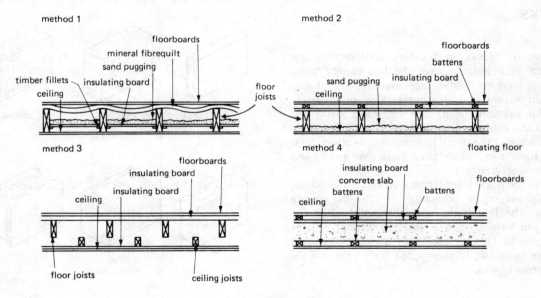

Figure 11.47

12
BLOCKWORK AND PARTITION WALLS

The essential purpose of partition walls is to divide buildings into suitably sized rooms, halls, etc. but they may also be required to support floor and ceiling joists. They can thus be described as loadbearing or non-loadbearing, and this factor may decide the type and thickness of the materials used for construction. For example, non-loadbearing walls may be erected using aerated concrete blocks only 60 mm thick, depending on the length and height whereas, if loads are to to superimposed, the minimum thickness for this type of block is 75 mm.

BLOCKS

The use of blocks in building construction has increased in popularity over the last twenty years since they have many advantages over bricks. Probably the main advantage is the increase in productivity since, for example, laying one 450 x 225 x 100 mm block is the equivalent of six bricks; and a 200 mm-wide block of the same size is equal to twelve bricks. Blocks have many other advantages over bricks and these are described later.

Blocks are made from autoclaved aerated concrete, lightweight-aggregate concrete or dense-aggregate concrete, the first being invariably solid while the others can be described as solid, hollow or cellular (see figure 12.1). Clay blocks are also manufactured but have been largely superseded for partition walls by concrete blocks.

Definition

A block is defined by BS 6073 as a walling unit exceeding in length, width or height the dimensions specified for bricks in BS 3921. The height of the block must not exceed either its length or six times its thickness.

Types (as given in BS 6073)

Solid

A block which contains no formed holes or cavities other than those inherent in the material.

Hollow

A block which has one or more formed holes or cavities which pass through the block.

Cellular

A block which has one or more formed holes or cavities which do not wholly pass through the block.

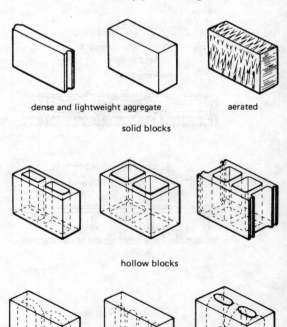

dense and lightweight aggregate   aerated

solid blocks

hollow blocks

cellular blocks

Figure 12.1 Solid, hollow and cellular blocks

Note Cavities in the wider cellular blocks may extend to the upper surface to provide small hand-holds for easy lifting. Cellular blocks are always laid holes downwards to provide a flat surface for the next spread of mortar.

Block Density

This is calculated by dividing the mass (weight) of a block by the overall volume, including holes and cavities.

Block Sizes

These are given in table 12.1. Blocks 75 mm thick and over may be used in general building work both above and below ground level for the construction of loadbearing and non-loadbearing walls. They may also be used for the inner leaf of external cavity walls provided that the *U*-value complies or can be made to comply with the requirements of the Building Regulations; and for the external leaf of cavity walls if suitably clad, rendered or tanked. Blocks less than 75 mm in thickness are intended for the construction of internal non-loadbearing walls only.

Aerated Concrete Blocks

These are probably the most popular blocks used in the construction of dwellings, mainly because when used for the inner leaf of an unventilated cavity wall they provide when plastered the required degree of thermal insulation and no other insulating material need be incorporated.

They are manufactured from cement, sand and pulverised fuel ash, to which an aerating agent, namely aluminium powder in suspension, is added.

These materials are thoroughly mixed and then poured into moulds varying in size between 1 and 4 m^3. The moulds are only partly filled because the materials swell when placed in the stiffening chambers; this is because the aluminium powder reacts with the free lime in the cement to release hydrogen, which is rapidly replaced by air, producing a light, micro-cellular structure. After stripping them from the moulds the cakes, as they are known at this stage, are cut into the required sizes by a series of wires, after which the blocks are high-pressure steam cured in an autoclave. The whole process from start to completion of curing takes about 24 hours.

The advantages of using this type of block are

(1) they have nearly three times the thermal insulation value of lightweight-aggregate blocks and six times that of brick
(2) the life of internal decorations is prolonged by the reduction of condensation on the surface
(3) in comparison with bricks, one-third the volume of mortar is required
(4) 100 mm blocks are a one-hand lift, thus productivity is increased
(5) they are easily cut, sawn or chased to suit any situation
(6) fixings can be directly made; there is no necessity for drilling and plugging
(7) they are completely frost resistant and are thus suitable for below d.p.c. or external walls when rendered
(8) the surface is ideal for plastering.

Lightweight-aggregate Concrete Blocks

These are manufactured from such materials as expanded vermiculite, clay and shales, foamed slag, sintered pulverised fuel ash, etc., the density being

Table 12.1 BS 6073 Part 2 1981: Work Sizes of Blocks

Thickness (mm)		60	75	90	100	115	125	140	150	175	190	200	215	225	250
Length (mm)	Height (mm)														
390	190		*	*	*	*		*	*		*	*			
440	140		*	*	*			*	*		*	*		*	
440	190		*	*	*			*	*		*		*		
440	215		*	*	*	*	*	*	*	*	*	*	*	*	*
440	290		*	*	*			*	*		*	*	*		
590	140			*	*	*		*	*		*	*	*		
590	190			*	*	*		*	*		*	*	*		
590	215			*	*	*	*	*	*	*		*	*	*	

Note: The compressive strength of concrete blocks usually varies from 2.8 to 35 N/mm^2.

less than 1500 kg/m³. They are mostly considerably heavier than blocks of aerated concrete and, while the degree of thermal insulation they offer is less, it is still better than that of bricks. They may be used for partition walls and in the construction of the internal leaf of cavity walls provided that the wall is adequately insulated against heat loss.

Dense Concrete Blocks

The density of these blocks is above 1500 kg/m³ and they have a higher compressive strength than those previously discussed. Dense concrete blocks are suitable for internal, external or structural walls and fair-faced blocks of this type present a good appearance if carefully laid and jointed. Specials are made by certain manufacturers such as Forticrete Ltd: they include quoin blocks, which immediately give half bond in either direction at return quoins; half and full-length cavity closers, which save on cutting and make a stronger job at reveals to openings; half blocks, which are very useful at stopped ends; and other useful shapes (figure 12.2).

CONSTRUCTION OF BLOCK WALLS

Internal Leaf of Cavity Walls

The quoins are set up, gauged and plumbed as with brickwork before running in to the line. The position of partition walls must be carefully measured out and provision made for tying these in, which is carried out using one of the following three methods:

1. Erecting inner leaf and partitions as work proceeds. The partition wall is squared off the inner leaf of the cavity and 'run out' say three blocks. As the inner leaf is erected the partition is racked back, tying in every other course. Where this method is used it is recommended that the area between the corners should be filled in within 2 hours (see figures 12.4 and 12.5).

2. Leaving indents (see figure 12.3).
3. Building in metal ties (see figure 4.9, page 25).

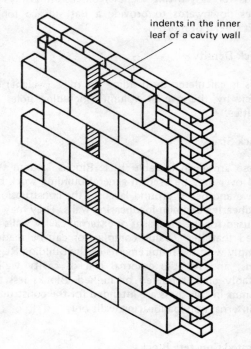

indents in the inner leaf of a cavity wall

Figure 12.3

Cutting, and Tying into a Chase (figure 12.6)

It is stated in the Building Regulations that no vertical chase must exceed one-third of the wall thickness and no horizontal chase must exceed one-sixth. Thus the wall to be chased for tying into must be at least 200 mm thick for this method to be considered.

Partition Walls

Non-loadbearing walls are usually built off the over-site concrete, which should be thickened below this point to sustain the loading. Loadbearing partitions should be erected from a concrete foundation taken to the same depth as the rest of the substructure. Assuming that the work to d.p.c. height is completed, the partition wall should be dry-bonded to determine the best bonding arrangement. A piece of 10 mm ply should be placed between each block and the next to allow for the mortar joints.

Two methods may be used from this point

(1) build up the corners as with brickwork and run in to the line
(2) fix a series of profiles at return corners and stopped ends and mark on the gauge (figure 12.6).

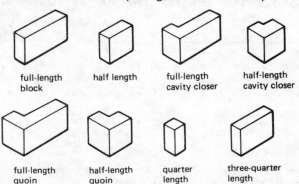

full-length block   half length   full-length cavity closer   half-length cavity closer

full-length quoin   half-length quoin   quarter length   three-quarter length

Figure 12.2

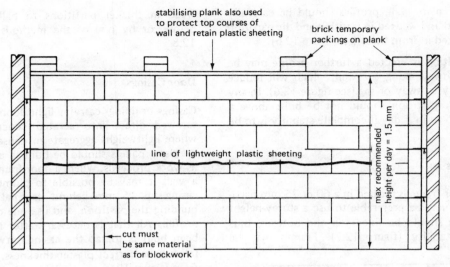

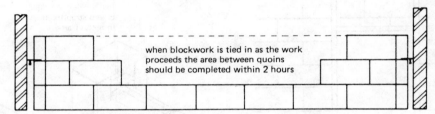

Figure 12.4 Building blockwork partition walls

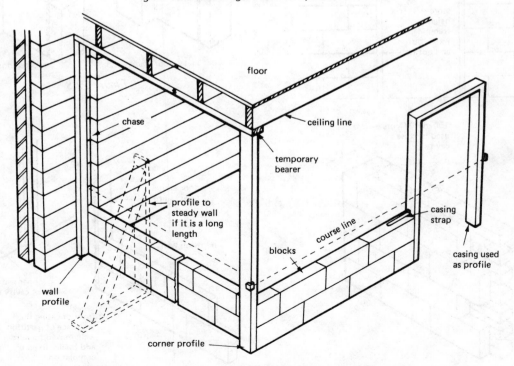

Figure 12.5 Building blockwork partition walls

Figure 12.6 Blockwork partition walls erected with profiles

At the cavity wall, profiles should be carefully plumbed and wedged in position and the partition walls lined in from these (see figure 12.6).

As the partition is erected a further profile may be required midway along the wall, which will reduce any tendency to sway or sag (see figure 12.6). In any case a block partition should not be built above a height of 1.5 m in 1 day if complete stability is to be achieved.

Door Frames

Where a door frame is located in a 60 or 75 mm partition wall it may be preferable to use a storey-height frame which should be fixed top and bottom, thus increasing stability (figure 12.7). Frames may be

fixed in thicker partitions to pallets as described below, or by one of the methods shown in figure 12.8.

Door Casings

Casings or linings carrying light doors can be nailed or screwed direct into aerated concrete blocks but, where lightweight aggregate or dense concrete blocks are used, pallets should be built in every two courses as work proceeds. Where a door frame or casing abuts a wall it may be possible to plumb and fasten the casing up 15 mm behind its final position before building the partition, and use one of the styles as a profile. All that is necessary after the partition has been built is to tap the casing outwards as required to give the correct plaster thickness, plumb it up and fix it (figure 12.9).

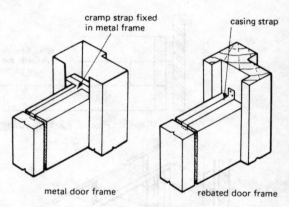

Figure 12.8

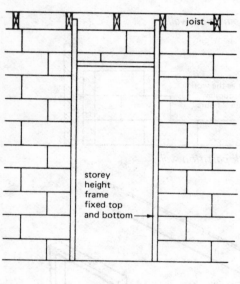

Figure 12.7

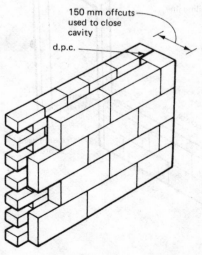

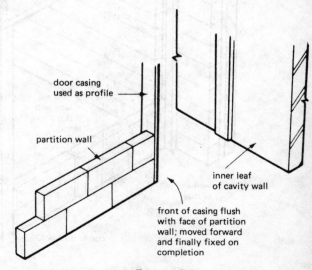

Figure 12.9

Partitions on Upper Floors

Where it is intended to build a block partition off a timber floor, as may be the case where two bedrooms are located directly above a long lounge, deflection must be limited either by inserting a joist of suitable strength directly below the partition or building in a BS beam between the joists (see figures 12.10 and 12.11).

When a block partition is to be built off a concrete floor the floor should be well hacked to form a key.

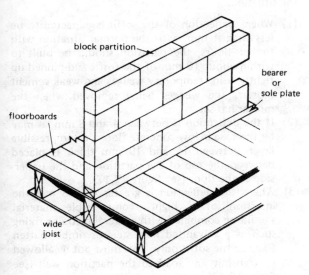

Figure 12.10

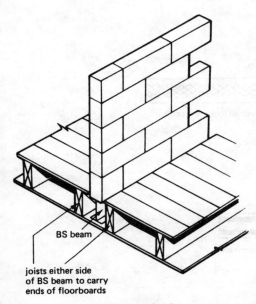

Figure 12.11

Bond

The most commonly used blocks are 450 mm long and 225 high, with varying widths, and it is to these blocks that the following information applies. Using quarter bond saves on cutting but a stronger job is produced when half bond is used. If specials are not available half bond can be achieved in two ways as follows.

(1) A block is cut into four equal pieces, each of which is used on consecutive courses in opposite directions next to the quoin (figure 12.12).
(2) A block is cut into two pieces, 300 mm and 150 mm in length. The 300 mm piece is used to start each course from the quoin and the 150 mm

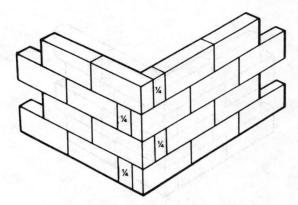

Figure 12.12 Method 1: half bond

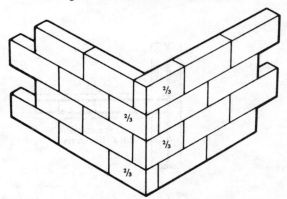

Figure 12.13 Method 2: half bond

piece is used to close the cavity at internal reveals to jambs of doors and windows (figures 12.7 and 12.13).

The use of specials and quarter bond are shown in figures 12.14 and 12.15. Where it is considered necessary to strengthen a partition wall without resorting

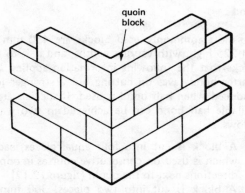

Figure 12.14 Use of special quoin blocks: half bond

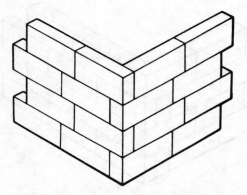

Figure 12.15 Use of quarter bond

to the use of attached piers, wider or denser blocks, longitudinal reinforcement can be built into alternate horizontal joints in the form of hoop iron or expanded metal, etc. (figure 12.16).

Treatment at the Top of Partitions

Where possible it is preferable to provide lateral stability by building the partition up between the joists and butting the ceiling up to either side.

If, however, the partition wall is erected later the treatment required below the soffit will depend on the situation.

(1) Where deflection of the soffit is expected to be less than 4 mm, as is the normal situation with timber floors, the partition should be built to within about 10 mm of the soffit and pinned up tightly using temporary wedges and weak cement mortar, the wedges being removed when the mortar has set.
(2) If the deflection is between 4 and 8 mm, as may be the case below concrete floors, a compressible layer between 15 and 19 mm thick is placed between the top of the partition and the underside of the floor (see figure 12.17).
(3) Where the deflection is greater than 8 mm the insertion of a highly compressible material should be combined with a lateral restraint fixing such as a galvanised mild steel or timber batten fixed to the soffit only. This must not be allowed to transmit any load to the partition wall (see figure 12.18).

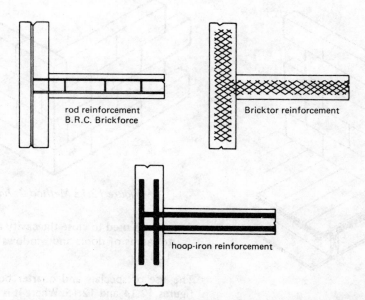

rod reinforcement
B.R.C. Brickforce

Bricktor reinforcement

hoop-iron reinforcement

Figure 12.16 Longitudinal reinforcement of block partition walls (plan views)

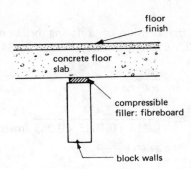

Figure 12.17 Block partition walls below solid floors

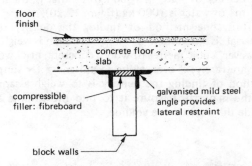

Figure 12.18 Block partition walls below solid floors

Precautions when Blocklaying

(1) Keep blocks dry before laying by stacking clear of the ground, on edge under cover. If, for example, aerated concrete blocks are laid in a wet condition, they will shrink as they dry out, resulting in vertical cracking of the completed walling.

(2) Use mortar slightly weaker than the blocks in order that any cracking is confined to the joints rather than to the blocks.

(3) Do not build long uninterrupted lengths of blockwork. Construction joints should be placed every 6 m and filled with a flexible mastic compound.

(4) If the blockwork becomes wet during building it must be allowed to dry out before plastering to reduce the risk of shrinkage cracking.

(5) Loadbearing partitions should be built of blocks not less than 75 mm wide.

(6) Make certain that the blocks used are of the required quality (class A, B or C).

(7) Cover walls down at the end of the day with polythene sheeting, with a plank on top. Further weight down with bricks if necessary (see figure 12.4).

DENSITY

The term density has been used in conjunction with brickwork and blockwork and it is necessary for the student or apprentice to have an understanding of the meaning of this term.

When we refer to a material as being light or heavy we mean in relation to another material. For example, when we compare aerated concrete blocks with those manufactured from dense concrete, the latter are considered heavy but when dense concrete blocks are compared with lead they are light.

Thus, simply to describe a material as light or heavy is insufficient, and a more accurate method of comparison is necessary. This is provided by the term density, which means the mass per cubic metre. For example, the density of lead is 11,300 kg/m³, that is, the mass of 1 m³ of lead is 11,300 kg (figure 12.19).

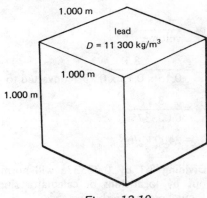

Figure 12.19

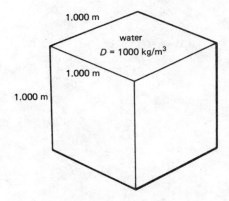

Figure 12.20

The density of water is 1000 kg/m^3, that is, the mass of 1 m^3 of water is 1000 kg (figure 12.20).

Thus, one method of finding the density of a material is to obtain a cubic metre and weigh it. However, cubic metres of wood, concrete, brickwork, etc. are rarely, if ever, to be found and a far simpler method of finding the density is to weigh a sample of the material, measure it to find the volume and divide the mass by the volume.

Note

When finding the density of a material, mass is always measured in kilograms and volume is always measured in cubic metres. It is therefore vital to carry out all calculations in kilograms and metres throughout. Thus

$$density = \frac{mass\ (kg)}{volume\ (m^3)}$$

Example 12.1

To find the density of a sample of concrete given a cube 150 x 150 x 150 mm and having a mass of 8.1 kg

$$density = \frac{mass}{volume}$$

$$= \frac{8.1}{0.15 \times 0.15 \times 0.15}\ (converted\ to\ metres)$$

$$= \frac{8.1}{0.003375}$$

$$= 2400\ kg/m^3$$

Note   Dividing 8.1 by 0.003375 will normally be carried out by logarithms or calculator since long division is cumbersome in this instance.

Example 12.2

To find the density of a facing brick measuring 215 x 102.5 x 65 and having a mass of 2.434 kg

$$density = \frac{mass}{volume}$$

$$= \frac{2.434}{0.215 \times 0.1025 \times 0.065}\ (metres)$$

$$= 1699\ kg/m^3$$

Thus, if the volume of the material can be found by calculation, its density too is easily obtained; but if the material has an irregular surface, such as a sample of rock or stone may have, and cannot be measured, the following method can be used.

Place a measuring cylinder below the outlet of a displacement can full of water and carefully lower the specimen into the can. The volume of water in the measuring cylinder will be equal to that of the object and can either be read off direct or determined by weighing and dividing the mass of the water by its density (1000 kg/m^3). Next weigh the specimen and find the density as before

$$density = \frac{mass}{volume}$$

Note   The specimen should be saturated before immersion.

MULTIPLE CHOICE QUESTIONS

Select your options from the questions below, underline your selection, for example (b), and check your answers with those on page 136.

1. In Flemish Garden Wall bond, the number of stretchers following each header is:
   (a) 3
   (b) 2
   (c) 4
   (d) 1

2. The toothing method is normally used when walls are to be:
   (a) built higher
   (b) terminated
   (c) corbelled out
   (d) extended

3. To comply with current Building Regulations, the minimum cavity width in an external wall is:
   (a) 50 mm
   (b) 60 mm
   (c) 75 mm
   (d) 100 mm

4. The most likely result of building a partition wall of aerated, lightweight concrete blocks in a wet condition is:
   (a) the wall will have a low bearing capacity
   (b) shrinkage cracking on drying
   (c) delayed setting of the cement mortar
   (d) expansion as the blockwork dries out

5. When block indents are formed to receive a tee-junction wall, the number of courses per indent is:
   (a) 2
   (b) 3
   (c) 4
   (d) 5

6. In timber ground floor construction, the d.p.c. should be bedded:
   (a) on the oversite concrete under the sleeper wall
   (b) on top of the sleeper wall under the wallplate
   (c) on top of the wallplate below the joists
   (d) not less than 150 mm above external ground level

7. When wall ties are built in next to unbonded jambs, the maximum vertical spacing must be:
   (a) 300 mm
   (b) 450 mm
   (c) 600 mm
   (d) 900 mm

8. Broken bond occurs in a wall when:
   (a) the general pattern of the bond cannot be kept
   (b) one end starts with a header and the other with a stretcher
   (c) English bond is combined with Flemish bond
   (d) diagonal cracking occurs through the perpends

9. The maximum depth for a vertical chase formed in the face of a wall is equal to the wall thickness divided by:
   (a) 2
   (b) 3
   (c) 4
   (d) 6

10. Which of the following would be the most suitable type of brick for use in forming thresholds?
    (a) Calcium silicate
    (b) Red rubber
    (c) Concrete
    (d) Engineering

11. On certain types of recessed reveals, the cut brick placed next to the small bevelled bat is a:
    (a) mitred bat
    (b) bevelled closer
    (c) king closer
    (d) quarter bat

12. The purpose of the hooked ends on reinforcement bars in concrete lintels is to:
    (a) resist compression
    (b) prevent the rods slipping
    (c) combat tension
    (d) increase rod strength

13. The leanest permissible mix of concrete for a concrete lintel is:
    (a) 1:1:4 cement:sand:stone
    (b) 1:2:4 cement:sand:stone
    (c) 1:3:6 cement:sand:stone
    (d) 1:2:8 cement:sand:stone

14. A rough arch consists of:
    (a) uncut bricks with wedge-shaped joints
    (b) cut bricks with parallel joints
    (c) uncut bricks with parallel joints
    (d) cut bricks and wedge-shaped joints

15. The material used to form the vertical joint between the tiled surround and the front edge of the fireback is:
    (a) cement mortar
    (b) fibreglass rope
    (c) tarred rope
    (d) lime mortar

16. When easing an arch centre it is advisable to:
    (a) remove the nails
    (b) take out the props
    (c) loosen the wedges
    (d) tap it gently with a hammer

17. The minimum thickness of a constructional hearth is:
    (a) 100 mm
    (b) 125 mm
    (c) 150 mm
    (d) 200 mm

18. Moisture is prevented from entering buildings at door and window jambs by:
    (a) mastic pointing
    (b) extra wall ties
    (c) cavity walling
    (d) vertical d.p.c.s.

19. A fender wall is built to support the:
    (a) jambs and constructional hearth
    (b) floor joists and constructional hearth
    (c) oversite concrete and floor joists
    (d) floor joists and jambs

20. English Garden wall bond consists of:
    (a) alternate courses of headers and stretchers
    (b) 3 stretchers to one header in each course
    (c) 3 courses of stretchers to one of headers
    (d) 3 courses of headers to one of stretchers

21. Wooden fixings are normally built into door and window jambs. These are called:
    (a) joggles
    (b) pallets
    (c) wedges
    (d) plugs

22. When a curved wall is to be built to a radius of 1.5 m, the most suitable bond would be:
    (a) stretcher
    (b) header
    (c) English
    (d) Flemish

23. In English and Flemish bond, the quoin header is normally followed by a:
    (a) stretcher
    (b) header
    (c) closer
    (d) half-bat

24. The reinforcing bars in a simply-supported lintel should be:
    (a) near the bottom
    (b) at the centre
    (c) near the top
    (d) at both ends

25. The amount of tolerance for a trammel pivot rod should be:
    (a) 2–3 mm
    (b) 5–6 mm
    (c) 8–10 mm
    (d) 10–12 mm

26. The purpose of a turning piece is to:
    (a) set out circular work
    (b) assist in building a bulls-eye
    (c) provide a check on curved walling
    (d) provide temporary support for certain arches

27. According to Building Regulations, the minimum thickness of separating walls is:
    (a) 100 mm
    (b) 150 mm
    (c) 200 mm
    (d) 190 mm

28. The minimum permissible distance from the top of the oversite concrete to the underside of the wallplate is:
    (a) 50 mm
    (b) 75 mm
    (c) 100 mm
    (d) 150 mm

29. The height of an isolated brick should not exceed:
    (a) six times the least dimension
    (b) four times the greatest dimension
    (c) eight times the least dimension
    (d) six times the greatest dimension

30. The sloping abutment supporting a segmental arch is called the:
    (a) springer
    (b) striking point
    (c) skewback
    (d) hauches

31. A purpose-made obtuse squint brick has a header face of:
    (a) 56 mm
    (b) 75 mm
    (c) 102 mm
    (d) 168 mm

32. A concrete lintel cast in formwork over an opening is known as:
    (a) precast
    (b) pre-stressed
    (c) cast *in situ*
    (d) purpose-made

33. If the angle between two radii of a sector of a circle is 40°, the area of the sector may be found from:
    (a) $\dfrac{2\pi r}{9}$

    (b) $\dfrac{\pi r}{9}$

    (c) $\dfrac{\pi r^2}{9}$

    (d) $\dfrac{2\pi r^2}{9}$

34. When broken bond occurs in straight lengths of walling, the smallest permissible bat is a:
    (a) three quarter
    (b) half
    (c) quarter
    (d) full brick

35. The term 'seating' refers to the:
    (a) underside of a concrete threshold
    (b) projection provided for brick sills
    (c) end bricks on a brick threshold
    (d) raised ends on a stone sill

36. Wall ties in straight lengths of cavity walling should be placed:

	horizontally	vertically
(a)	900	300
(b)	450	900
(c)	900	450
(d)	900	600

37. The usual thickness for the joints between the voussoirs in a gauged arch is:
    (a) 3 mm
    (b) 5 mm
    (c) 10 mm
    (d) 15 mm

38. The horizontal distance by which one brick projects beyond the vertical joint in the course below is called the:
    (a) bond
    (b) lap
    (c) half bond
    (d) quarter bond

39. Walls provide better thermal insulation when the *U*-value is:
    (a) 0.35 W/m²/°C
    (b) 1.00 W/m²/°C
    (c) 1.35 W/m²/°C
    (d) 2.35 W/m²/°C

40. The voussoirs for an axed, segmental arch are set out on the:
    (a) centre line
    (b) springing line
    (c) intrados
    (d) extrados

ANSWERS TO MULTIPLE CHOICE QUESTIONS

1. (a)
2. (d)
3. (a)
4. (b)
5. (c)
6. (b)
7. (a)
8. (a)
9. (b)
10. (d)
11. (c)
12. (b)
13. (b)
14. (a)
15. (b)
16. (c)
17. (b)
18. (d)
19. (b)
20. (c)
21. (b)
22. (b)
23. (c)
24. (a)
25. (a)
26. (d)
27. (d)
28. (b)
29. (a)
30. (c)
31. (a)
32. (c)
33. (c)
34. (b)
35. (d)
36. (c)
37. (a)
38. (b)
39. (a)
40. (d)

INDEX